Plant-Based Genetic Tools for Biofuels Production

Edited by

Daniela Defavari do Nascimento

*Laboratory of Biotechnology, Faculdade de Tecnologia de Piracicaba
"Dep. Roque Trevisan" (FATEC Piracicaba), Centro Estadual de Educação
Tecnológica Paula Souza (CEETEPS), Piracicaba-SP, Brazil*

&

William A. Pickering

*Faculdade de Tecnologia de Piracicaba "Dep. Roque Trevisan"
(FATEC Piracicaba), Centro Estadual de Educação Tecnológica Paula Souza
(CEETEPS), Piracicaba-SP, Brazil*

Plant-Based Genetic Tools for Biofuels Production

Editors: Daniela Defavari do Nascimento, and William A. Pickering

eISBN (Online): 978-1-68108-461-9

ISBN (Print): 978-1-68108-462-6

General:

1. Any dispute or claim arising out of or in connection with this License Agreement or the Work (including non-contractual disputes or claims) will be governed by and construed in accordance with the laws of the U.A.E. as applied in the Emirate of Dubai. Each party agrees that the courts of the Emirate of Dubai shall have exclusive jurisdiction to settle any dispute or claim arising out of or in connection with this License Agreement or the Work (including non-contractual disputes or claims).
2. Your rights under this License Agreement will automatically terminate without notice and without the need for a court order if at any point you breach any terms of this License Agreement. In no event will any delay or failure by Bentham Science Publishers in enforcing your compliance with this License Agreement constitute a waiver of any of its rights.
3. You acknowledge that you have read this License Agreement, and agree to be bound by its terms and conditions. To the extent that any other terms and conditions presented on any website of Bentham Science Publishers conflict with, or are inconsistent with, the terms and conditions set out in this License Agreement, you acknowledge that the terms and conditions set out in this License Agreement shall prevail.

Bentham Science Publishers Ltd.
Executive Suite Y - 2
PO Box 7917, Saif Zone
Sharjah, U.A.E.
Email: subscriptions@benthamscience.org

**BENTHAM
SCIENCE**

CONTENTS

PREFACE

The growing dependency on energy derived from depleting supplies of fossil fuels no longer has a future. The combustion of oil as an energy source is one of the biggest causes of air pollution, due to the releasing of CO_2 in the environment. Solar energy, on the other hand, when captured by plants through photosynthesis, promotes the assimilation of CO_2 and results in the opposite of the "greenhouse effect". Increased demand for fuels of vegetable origin, in addition to their environmental appeal, has made these fuels an appealing alternative and increased their production possibilities. This is where the need has arisen for the application of different biotechnological techniques, such as viable biomass production, development of cell wall degrading tools, and efficient fermentation technology.

The choice of biomass species is a key step in the production of biofuels with high oil and carbohydrate content. There is a great potential for the use of different kinds of crop residues for the production of bioenergy. After sugarcane juice is extracted, the remaining biomass (bagasse), rich in cellulose, can be used for second generation biofuel. A similar process can also be performed using corn stover, municipal solid wastes, and forestry residues. Evidence suggests that the current production of biofuels is at less than capacity. This is considered to be due to a confluence of factors, including high feedstock and processing costs, regulatory frameworks, risk avoidance, and limits to the amount of biofuel that can be blended with conventional fuels in major markets.

The use of cell suspension, somatic embryogenesis, gametic embryogenesis, and protoplast fusion, presented in chapters 1 and 2, give some insight into the ways that these techniques can be used to produce renewable fuels. Biomass feedstock production may benefit from the identification and characterization of key proteins involved, for example, in the biosynthesis of cell wall components and oil bodies. In this context, as seen in chapter 6, the exploration of subcellular proteomes also shows great promise in the characterization of new protein families and regulation mechanisms for improved biofuel crops. As discussed in chapter 4, new plant breeding techniques comprise a group of methods that offer the possibility of performing precise editing, replacement, or insertion of genes, targeting specific genomic regions without the use of any selectable marker. These features minimize the probability of undesirable random gene disruption, thereby providing interesting alternative tools for genetic transformation. The new tools are more predictable and less prone to position effects than are conventional methods.

All living organisms are constantly exposed to a variety of DNA-damaging agents. Chapter 3 shows that mutations in DNA are frequently observed in several diseases, a factor which is reflected in metabolic changes of great importance in biotechnology and biofuels production. It is seen in Chapter 5 that viruses constitute very powerful targets or tools for plant biotechnology applications. Plant viruses can be efficient vehicles for heterologous gene expression in plants used as biofactories or biofuels sources. Biotechnology is playing a major role in new advances in the fermentation of different substrates and in the production, not only of ethanol, but of biodiesel, butanol, and many other biofuels, as described in chapter 9. Genomic resources and bioinformatic tools are available for plant species with bioenergy and biofuels potential, and these are presented in chapter 7.

Plant biomass is the main feedstock for biofuels production. Efforts to maximize yield per unit of production area are of crucial importance in meeting the rising demand for renewable energy sources. In chapter 8, a broad overview is given of the factors influencing biomass yield, of advances in cultivation technologies, and of the relation of these conditions to the physiology of energy crops. The chapter also presents innovative technologies that can support management decisions, focusing on sugarcane as a model for bioenergy crops. Microalgae biomass has also been described by several authors as an alternative with great potential for accomplishing the goal of the replacement of diesel by biodiesel, while at the same time not competing with fertile land useable for food production. However, the technology must still overcome a number of obstacles in order to be widely deployed. The advances and challenges of the technologies used, including procedures for obtaining biomass, are presented in chapter 10.

As is clear from the above discussion, this eBook is aimed at addressing sustainable biotechnological techniques that have been applied in the search for significant increases in biofuel productivity, without affecting food production. It is expected that the lessons from the sustainability criteria applied to first generation biofuels will be incorporated into advanced biofuels production, not necessarily focusing on the type of supply, but instead on existing local demands and domestic development strategies.

Daniela Defavari do Nascimento
Laboratory of Biotechnology,
Faculdade de Tecnologia de Piracicaba "Dep. Roque Trevisan" (FATEC Piracicaba),
Centro Estadual de Educação Tecnológica Paula Souza (CEETEPS),
Piracicaba-SP,
Brazil

&

William A. Pickering
Faculdade de Tecnologia de Piracicaba "Dep. RoqueTrevisan" (FATEC Piracicaba),
Centro Estadual de Educação Tecnológica Paula Souza (CEETEPS),
Piracicaba-SP,
Brazil

List of Contributors

Alessandro Antonio Orelli	Faculty of Technology of Piracicaba, Paula Souza Center, Piracicaba, Brazil
Andrea L. Venturuzzi	Institute of Biotechnology, Center of Veterinary and Agricultural Research – IB – CICVyA – INTA, Argentina
Bruna Marques dos Santos	Department of Technology, Paulo State University, Jaboticabal, Brazil
Camila Caldana	Laboratório Nacional de Ciência e Tecnologia do Bioetanol (CTBE), Centro Nacional de Pesquisa em Energia e Materiais (CNPEM), Campinas, Brazil
Carlos Alberto Labate	Department of Genetics, Escola Superior de Agricultura Luiz de Queiroz (ESALQ), University of São Paulo (USP), Piracicaba, Brazil
Daniela Defavari do Nascimento	Faculty of Technology of Piracicaba, Paula Souza Center, Piracicaba, Brazil
Daniela Kubiak de Salvatierra	Biofábrica Misiones S.A., Posadas, Misiones, Argentina
Diego Zavallo	Institute of Biotechnology, Center of Veterinary and Agricultural Research – IB – CICVyA – INTA, Argentina
Edwin Antônio Gutierrez Rodriguez	Department of Plant Production, São Paulo State University, Jaboticabal, Brazil
Evandro Novaes Escola de Agronomia	Universidade Federal de Goiás, Goiânia, Brazil
Fabio V. Scarpare	Faculdade de Engenharia Mecânica (FEM), Universidade Estadual de Campinas (UNICAMP), Campinas, SP, Brazil
Felipe Augusto Godoy	Department of Biochemistry, Institute of Chemistry, University of Sao Paulo, São Paulo, Brazil
Fernanda Salvato	Department of Plant Biology, Institute of Biology, University of Campinas, Campinas, Brazil
Gabriela Conti	Institute of Biotechnology, Center of Veterinary and Agricultural Research – IB – CICVyA – INTA, Argentina
Gisele G. Bortoleto	Faculty of Technology of Piracicaba, Paula Souza Center, Piracicaba, Brazil
Henrique L. de Miranda	Faculty of Technology of Piracicaba, Paula Souza Center, Piracicaba, Brazil
Humberto J. Debat	Institute of Plant Pathology, Center of Agricultural Research – IPAVE – CIAP – INTA, Argentina
João L. N. Carvalho	Laboratório Nacional de Ciência e Tecnologia do Bioctanol (CTBE), Centro Nacional de Pesquisa em Energia e Materiais (CNPEM), Campinas, Brazil
Jose Antonio Cabral	Biofábrica Misiones S.A., Posadas, Misiones, Argentina

Lucia Mattiello	Departamento de Genética, Evolução e Bioagentes, Laboratório de Genoma Funcional, Universidade Estadual de Campinas (UNICAMP), Campinas, Brazil
Maria J. Calderan-Rodrigues	Laboratório Nacional de Ciência e Tecnologia do Bioetanol (CTBE), Centro Nacional de Pesquisa em Energia e Materiais (CNPEM), Campinas, Brazil
Mariane B. Sobreiro	Escola de Agronomia, Universidade Federal de Goiás, Goiânia, Brazil
Marília Gabriela de Santana Costa	Department of Technology, São Paulo State University, Jaboticabal, Brazil
Marina C. M. Martins	Laboratório Nacional de Ciência e Tecnologia do Bioetanol (CTBE), Centro Nacional de Pesquisa em Energia e Materiais (CNPEM), Campinas, Brazil
Marines Marli Gniech Karasawa	Dipartimento di Scienze Agrarie e Forestali, Università degli Studi di Palermo, Palermo, Italy
Mateus Prates Mori	Department of Biochemistry, Institute of Chemistry, University of Sao Paulo, São Paulo, Brazil
Rodrigo H. de Campos	Faculty of Technology of Piracicaba, Paula Souza Center, Piracicaba, Brazil
Sabrina D. Soares	Escola de Agronomia, Universidade Federal de Goiás, Goiânia, Brazil
Vanessa C. Araújo	Escola de Agronomia, Universidade Federal de Goiás, Goiânia, Brazil
Verónica C. Delfosse	Institute of Biotechnology, Center of Veterinary and Agricultural Research – IB – CICVyA – INTA, Argentina
William A. Pickering	Faculty of Technology of Piracicaba, Paula Souza Center, Piracicaba, Brazil
Yamila C. Agrofoglio	Institute of Biotechnology, Center of Veterinary and Agricultural Research – IB – CICVyA – INTA, Argentina

CHAPTER 1

Biotechnology and its Impact on Vegetative Propagation of Plant Species

Guillermo R. Salvatierra[*], **Daniela Kubiak de Salvatierra** and **Jose Antonio Cabral**

Biofábrica Misiones S.A., Posadas, Misiones, Argentina

Abstract: The application of biotechnology has had great impact on the agricultural sciences. Micropropagation, in particular, is one of the biotechnological methods whose major achievements have contributed to the development of agriculture in Northeast Argentina, and it is used in the mass production of aromatic, medicinal, fruit, ornamental, and forest plant species. It is normally applied to certified cultivars with good productive performance, providing significant development to the sector. Micropropagation also provides significant production and economic benefits, and an unprecedented environmental contribution.

Keywords: Biotechnology, Tissue culture, Vegetal micropropagation.

BIOTECHNOLOGY

After its first application in the cattle sector, the term biotechnology evolved in association with industrial fermentation. In 1961, a Swedish microbiologist defined it as the industrial production of goods and services through the use of organism systems or biological processes. The use of microorganisms thus became reflected in the concept. Yeasts were used to allow fermentation processes in the production of wine and beer, and antibiotics were obtained from fungi. Insulin and vaccines against hepatitis B are also produced by microorganisms, encompassing what is called industrial biotechnology [1, 2].

The development of recombinant DNA technology in the 1970s allowed plants and animals to behave as new gene product bioreactors. This opened a new horizon of great impact on agricultural and animal sciences that would complement the advance and release of genome projects for several species, where countless coding sequences of interest were discriminated and categorized

[*] **Corresponding author Guillermo R. Salvatierra:** Biofábrica Misiones S.A., Posadas, Misiones, Argentina; Tel: +54 9376 4268922; E-mail: guillermosalvatierra7@gmail.com

Daniela Defavari do Nascimento, & William A. Pickering (Eds.)

by functionality. This generated possibilities for the construction of different transformation vectors [2].

Recently, biotechnology has been defined as the application of science and technology to living organisms, as well as to the parts, products, and models thereof, so as to alter living or non-living materials for the production of knowledge, goods, and services. This general framework allows us to include or add various techniques such as cell/tissue culture, biological pest control, biological supply production (pesticides, fertilizers, and fungicides of biological origin), genomics, and gene expression profiling, as well as techniques that allow direct and targeted modification of DNA, genetic engineering, and the introduction of new features in natural genomic sequences. A new field is thus opened with an unprecedented production potential, one that is especially relevant for the agricultural and forestry sector due to the characteristics and qualities of plants used in this type of study [2 - 4].

PLANT BIOTECHNOLOGY

Within what is called sustainable agriculture, the social, ecological, and economic aspects are crucial and prerogative. According to various researchers and economists, it is estimated that the world population will increase by about a third between 2009 and 2050 [5, 6]. This translates into an increase of 2300 million people, with growth occurring mainly in developing countries. Therefore, particularly in these countries, there will be a greater demand for food. To meet this demand there is a priority for road construction, increase of arable land, and improvement in performance and/or crop adaptation to marginal conditions. The first factor is insufficient for, and even detrimental to, the protection of natural environments [7]. However, the last factor is more desirable and points to South America as a great producer, as well as to its developmental potential for biotechnology and agricultural sciences.

The contributions of recombinant DNA technology, coupled with the progress of advanced genomics, formed what was previously called modern biotechnology [4], often supplemented by contributions from mass clonal propagation [2, 8]. In this context, companies such as Genentech [9], Monsanto [10], Syngenta [11], and Amgen [12], have been developing varieties of corn, tomatoes, and soybeans, among other species, and have improved various features such as herbicide tolerance, insect and virus resistance, and tolerance to abiotic stresses [2, 9, 12].

With the introduction of biotechnological techniques, new products, and new markets, a new economy has been generated, leading to greater production per unit area. This innovative concept of bioeconomy enables sustainable production with reduced costs, while improving product quality and the development of less aggressive environmental practices [13].

PLANT BIOTECHNOLOGY IN ARGENTINA

Argentina can be divided into five major distinct regions, whose soil and climatic characteristics determine their production profile: NEA (Northeast Argentina), NOA (Northwest Argentina), Cuyo, the Pampean region, and Patagonia [2]. As the country is an efficient and diverse producer of high-quality food due to its deep and rich soils, mild climate, adequate rainfall, and good access to maritime transport, it has great potential for the application of modern technologies in the value chain and in processes [6, 14].

Since 1996, many producers in Argentina have been steadily growing genetically modified plants (GMOs) [15]. In 2003, the Argentinian position in the world market was second among the eighteen countries that extensively cultivate GMOs, due to its fourteen million cultivated hectares. In 2007, some GMO varieties tolerant to insects and herbicides, such as soybean and maize, were released on the Argentinian market, and in 2009 cotton varieties were introduced. In 2012, twenty-four million hectares were used for GMO cultivation, ranking Argentina as third in the world in GMO use [14, 16].

The incorporation of biotechnology tools into agricultural production in the 1990s led producers from the perception of potential profits to the reality of actual earned profits. Currently, there is empirical support for the economic benefits, such as higher yields, of various species treated with biotech tools. The following results have been obtained: increased income; reduced production costs (reduced tillage, cheaper herbicides, fewer pesticide applications); agronomic benefits, such as synergistic complementarities with direct seeding; health benefits (reduced application of herbicides and insecticides); and environmental benefits that allow the incorporation of technologies having less environmental impact and promoting carbon sequestration [16 - 18].

In the past ten years, Argentina has had the highest agricultural growth in its history. The new technologies have allowed for a threefold increase in productivity and acreage, and have led to a sevenfold increase in productivity. The highest impact factor for this leap in Argentinian productive agribusiness was change. It has been estimated that two thirds of the increase was due to the incorporation of new technologies [19, 20].

In productive agricultural regions outside the Pampas, there is a wide range of ecological conditions and a variety of crops. These conditions demand management policies that favor competitiveness, such as public policies for the generation and transfer of technology, the implementation of sanitary and phytosanitary systems, and access to credit for agricultural improvements [14].

These policies can be promoted in various regions, taking advantage of advances in biotechnology such as genetic engineering of plants for resistance to various biotic and abiotic adverse factors, gene silencing of a gene of interest, and synthesis of specific proteins [14]. Among all biotechnological methods, tissue/cell culture has resulted in a great positive impact on plant biotechnology. Meristem culture, associated with thermotherapy, is an example of the production of virus-free plants and of the conservation of genetic material through the establishment of a germplasm bank by *in vitro* cultures. Various other techniques of tissue culture involving organogenesis and somatic embryogenesis can also be mentioned [21, 22]. These techniques interact synergistically with transgenesis technologies, permitting the incorporation of genes that provide resistance to fungal or bacterial diseases, and even to certain pests or adverse environmental factors. In addition, tissue/cell culture offers the possibility of quickly regenerating and propagating genetically modified plant tissue in sufficient quantity [23 - 25].

IMPACT ON VEGETATIVE PROPAGATION

As a productive area, NEA has a production profile that is distinct from the rest of the country's zones. It is comprised of five states: Chaco, Formosa, Corrientes, Entre Rios, and Misiones, and is characterized by abundant rainfall, high temperatures, and by being prone to heat waves. In this region there are two biotechnology centers which contribute to the productive sector: Biofábrica Misiones S.A [26]. and the Forestry and Agricultural Biotechnology Center of Chaco [27] (located in the states of Misiones and Chaco, respectively). There are also several universities and agricultural technology institutes dedicated to research and development of technologies. An example of their work is the draft genome of *Ilex paraguariensis*, which has been successfully completed [28].

The state of Misiones in particular is in a strategic position within the NEA and Mercosur. This state has the potential for large production of biomass and biodiversity, due to its heterogeneity of soils and its warm and humid tropical and temperate climate. Ample water availability in most of its territory naturally generates the development activities of the agro-industrial and livestock sectors. However, the state has a predominance of medium, small, and micro landholders, and has an economy that is mainly based on poorly technofied primary activities such as forestry, agriculture, and horticulture [29].

In Misiones, production areas have very specific dimensions, as this state has the lowest average area per producer in Argentina. In other words, the farms and profitable fields are much smaller in Misiones than in the rest of the country. According to the National Agricultural Census of 2008, approximately 80% of agricultural plots have less than fifty hectares [29], and these constitute 40% of the rural population [30]. Agricultural policy must therefore be oriented toward activities which require small land areas. An alternative would be to promote the grouping of small landholders into larger entities (*e.g.,* different producer cooperatives) [31 - 34].

For these reasons, the use of GMO plant species is not extensive in the state of Misiones, and it cannot be made larger due to the features of the producers and productive plots in this state [29]. Mixed systems, such as livestock-silvicola, agro-livestock-silvicola, and agro-silvicola have been established [31, 34 - 37]. In this case, diversified production has fundamental importance for the certification of cultivars in relation to their genetic purity, freedom from diseases, and high productivity. For these conditions and kinds of productive trait, there are particular biotechnological tools that help the producer. Micropropagation (clonal propagation carried out through *in vitro* tissue/cell culture) is one of the biotechnological methods which lead to major achievements, contributing to the development of agriculture in the NEA region. It is used in the mass production of horticulture, herbs, medicinal plants, fruit, and ornamental and forest crops, and it is normally applied to the certification of cultivars with good production performance [22, 25, 38].

Tissue culture is the *in vitro* aseptic culture of cells, tissues, organs, or whole plants, under controlled nutritional and environmental conditions. It is being used for large-scale plant multiplication and is an essential step for obtaining regenerated healthy plants (free from diseases), whether genetically homogeneous or genetically modified. Moreover, this technique can be also used in plant breeding programs for the production of secondary metabolites of interest [22, 23, 25, 39].

Micropropagation techniques can be classified based on their response in the culture media and their respective phytoregulators. They may undergo dedifferentiation accompanied by "tumor" growth, the product of which is an undifferentiated mass of cells called callus. Under appropriate conditions this process can generate somatic embryos or organ. It can also provide a morphogenetic response, generating organs (organogenesis) or embryos (somatic embryogenesis). The first response is called indirect embryogenesis or organogenesis (being mediated by a callus state), and the second response is called direct embryogenesis or organogenesis [22, 25, 38].

The commercial production of plants obtained with the help of micropropagation techniques has several advantages when compared to traditional propagation methods such as seeding, cutting, and grafting. Micropropagation techniques allow massive plant propagation, especially in cases where a particular species presents a difficulty for propagation with traditional methods, or is facing extinction. They are also useful when the goal is propagation in a short time, or obtaining better plants that are free of endogenous pathogens and are younger and more vigorous [21 - 23]. Moreover, micropropagation is an important technique for germplasm conservation. It is also used in plant breeding programs to introduce interesting agronomic characteristics into commercial cultivars. It is useful in the production of healthier plants, of synthetic seeds, and of new hybrids with good uniformity and constant production throughout the year [25].

The biofactory Biofábrica Misiones S.A. (BIOMISA) is a corporation whose major shareholder is the state of Misiones. Its vision is to be a leader in the efficient implementation of massive vegetative propagation technology, while adjusting biotechnology to the scope of the region's producers, who are mainly micro, small, and medium sized landholders. It can be defined as a productive company specialized in the vegetative propagation of plants from *in vitro* cultures, with a nursery that has the mission of acclimating the plantlets in *ex vitro* conditions. It offers a variety of quality biotechnological products that best suit the needs of producers and their respective production realities, and it also facilitates the logistics for the transportation of these materials.

In synergy with agricultural policies, since 2006 this biofactory has generated its own products using micropropagation techniques. Production is maintained for species of regional interest (Figs. **6**, **7**), such as *Eucalyptus grandis* (eucalyptus) (Fig. **1**), *Manihot esculenta* (cassava) (Fig. **5**), *Ananas comosus* (pineapple), *Musa* sp. (banana), *Saccharum officinarum* (sugarcane), various orchids, *Stevia rebaudiana* (stevia), and aromatic, ornamental (Fig. **2**), and medicinal plants, among other species (Fig. **6**). With the appropriate growing conditions for each explant type, plants can be induced to rapidly produce new shoots and, with the addition of phytoregulators, new roots. The new plants can then be placed in soil and grown in the normal manner, in order to maximize the characteristics of biotech products with guaranteed superior genetics, health, and quality. Regional agribusiness clusters can thus start their crops with appropriate biological material, allowing an increase in yield with better income. Recently, BIOMISA established field traits for new products, including *Pawlonia tomentosa* (kiri) and *Melia azederach* (paraiso) (Fig. **3**), and *Pennicetum* sp. (Fig. **4**). As a result, one million fruit plantlets (mainly banana and pineapple) have been shipped, and over three million plantlets are used in the industry (sugarcane and cassava). In addition, around four million plantlets have been produced in the aromatic category (stevia, mint, lemon verbena, sage, carqueja), as well as more than 100,000 orchid plantlets.

Fig. (1). *Eucalyptus* micropropagation. a) phase 1: explants placed on solid culture medium; b) phase 2: bud multiplication; c) the emerging shoots may be sliced off; d) phase 3: plantlet regeneration; e) plantlet rooted; f and g) phase 4: plantlets acclimated in greenhouse.

Fig. (2). *Heliconia* micropropagation. a) phase 1: explants placed on solid culture medium; b) phase 2: bud multiplication in liquid media; c) phase 2: bud multiplication in solid media; d) the emerging shoots may be sliced off; e) phase 3: plantlet regeneration; f) phase 4: plantlets acclimated in greenhouse.

Fig. (3). *Melia azederach* micropropagation. a) phase 1: explants placed on solid culture medium; b) phase 2: bud multiplication in solid media; c) phase 3: regenerated plantlets; d) phase 3: regenerated plantlet; e) phase 4: plantlets acclimated in greenhouse; f) phase 4: plantlet acclimated in greenhouse; g) plantlets ready for shipping; h) plantlet rustification in greenhouse.

Fig. (4). *Penicetum* sp. micropropagation. a) phase 1: explants placed on solid culture medium; b) phase 1: explants placed on solid culture medium, after few days; c) phase 2: bud multiplication in liquid media; d) phase 3: regenerated plantlets; e) phase 3: different types of regenerated plantlets; f) phase 4: plantlets acclimated in greenhouse; g) plantlets in greenhouse ready for shipping.

Fig. (5). *Manihot* sp.(cassava) micropropagation. a) phase 3: regenerated plantlets; b) phase 3: different types of regenerated plantlets; c) phase 4: acclimated plantlets, after few days; d) phase 4: plantlets   acclimated; e) rooted plantlets in greenhouse ready for transplanting and shipping; f) phase 4: plantlet acclimation in greenhouse.

Fig. (6). *Crocus* sp. (saffron) micropropagation (species in research and development). a) phase 2: multiplication in liquid media; b) phase 2: multiple microcorm developed in liquid media; c) phase 3: microcorm rooting; d) phase 3: microcorm shooting and microcorm development.

Fig. (7). Other species in research and development: *Pistacia vera* and *Ilex paraguariensis*. a) phase 1: *Pistacia vera* explants placed on solid culture medium; b) phase 1: *Ilex paraguariensis* explants placed on several solid culture media.

This process includes technology management, human resources training, consulting, and efforts to communicate and raise awareness among farmers and the general public about the benefits of the use of biotechnology and its products. Since 2008, it has created a social and productive chain allowing small holders to assume a productive role, in place of the subsistence production that was previously developed.

BIOTECHNOLOGY MICROPROPAGATION: FUTURE PERSPECTIVES

It is likely that one of the disadvantages of micropropagation, which reduces its impact on small producers, is the cost of producing plants through this method. The magnitude of its impact in the coming years will depend on the reduction of production costs, and on the interaction between education and research centers and the biotech companies that generate products at an efficient commercial scale. This interaction should be covered by policies that encourage the availability and use of biotechnological products by all, especially small and micro producers.

Plant biotechnology has the potential to develop and increase food production, parallel to additional benefits through biofortification, as with the production of GMOs such as golden rice. There are also other examples, such the BioCassava, the BT eggplant, the virus resistant potato, herbicide tolerant sugarcane, and insect resistant cultivars, among others, including those with increased sugar content. Thanks to micropropagation techniques, GMOs can be propagated in a short time. Plant genetic transformation is closely related to the pharming trend, due to the latter's ability to generate plants which produce various compounds such as vaccines and antibodies. Plant biotechnology provides important contributions toward mitigating the effects of climate change and toward the adaptation of crops to climate change, leading to more sustainable handling, with social benefits and the lowest possible environmental impact.

CONFLICT OF INTEREST

The authors confirm that this chapter content has no conflict of interest.

ACKNOWLEDGEMENTS

The authors would like to acknowledge the assistance of E. M. Escalante and the research and development team at Biofábrica Misiones S.A. (A. Dominguez, M. Trinidad, and R. Saleski), who provided both contributions and photos.

REFERENCES

[1]   Applied Molecular Genetics. Profesionales médicos: Biotecnología , [25th Nov 2015]; Available at: http://www.amgen.es/profesionales/biotecnologia

[2]   Vera D. Fruticultura & Diversificación. Biotecnología: Biotecnología y agricultura Alto Valle: Instituto Nacional de Tecnología Agropecuaria. 2009; 61: pp. 30-7.

[3]   Organisation for Economic Co-operation and development. Directorate for Science, Technology and Innovation Biotechnology policies Statistical Definition of Biotechnology , [20th Nov 2015]; Available at: http://www.oecd.org/sti/biotech/statisticaldefinitionofbiotechnology.htm

[4]   de Oliveira AM, Santos da Silva R, Barbosa MS. A biotecnologia aplicada ao melhoramento genético vegetal: controvérsias e discussões. Revista da Universidade Vale do Rio Verde 2012; 10(1): 339-61. [http://dx.doi.org/10.5892/ruvrv.2012.101.339361]

[5]   Food and Agriculture Organization of the United Nations. Foro de Expertos de Alto Nivel: La agricultura mundial en la perspectiva del año 2050 , [30th Nov 2015]; Available at: http://www.fao.org/fileadmin/templates/wsfs/docs/Issues_papers/Issues_papers_SP/La_agricultura_m undial.pdf

[6]   Kern M. Plant biotechnology: Perspectives for developing countries between 2002 and 2025. African J Food and Nutritional Sci 2002; 2(2): 39-46.

[7]   Food and Agriculture Organization of the United Nations. Towards a water and food secure future: Critical Perspectives for Policy-makers Food and Agriculture Organization of the United Nations: Rome 2015. Marseille: World Water Council 2015.

[8]   Kumar AP. Plant Biotechnology: Future perspectives. Def Sci J 2001; 51(4): 353-66. [http://dx.doi.org/10.14429/dsj.51.2249]

[9]   Genentech. Biotechnology Pioneer Genentech Turns 20: Five marketed medicines and more than $900 million a year in revenues later, Genentech remains a leader of the industry launched with its founding , [30th Nov 2015]; Available at: http://www.gene.com/media/press-releases/4824/1996-04-04/ biotechnology-pioneer-genentech-turns-20

[10]   Monsanto. Biotecnología para la agricultura , [30th Nov 2015]; Available at: http://www.monsanto. com/global/ar/productos/documents/biotecnologia-para-la-agricultura.pdf

[11]   Syngenta. Biotechnology , [30th Nov 2015]; Available at: http://www.syngenta.com/global/ corporate/en/products-and-innovation/research-and-development/biotechnology/pages/ biotechnology.aspx

[12]   Amgen. Historia , [30th Nov 2015]; Available at: http://www.amgen.es/acerca/historia

[13]   Pfau SF, Hagens JE, Dankbaar B, Smits AJ. Visions of sustainability in bioeconomy research. Sustainability 2014; 6: 1222-49. [http://dx.doi.org/10.3390/su6031222]

[14]   Schneider R, Verner D, Caballero JM, Miodosky M. Argentina Agriculture and Rural Development: Selected Issues - Report No 32763-AR. 1st ed., Washington DC: Banco Mundial 2006.

[15]   Willmitzer L. Plant biotechnology: output traits — the second generation of plant biotechnology products is gaining momentum. Curr Opin Biotechnol 1999; 10: 161-2. [http://dx.doi.org/10.1016/S0958-1669(99)80028-7]

[16]   Argenbio. 10 años - Difundiendo la biotecnología en Argentina , [1st Dec 2015]; Available at: http://www.argenbio.org/adc/uploads/fichas/Fichas10.pdf

[17]   Traxler G. The Economic Impacts of Biotechnology-Based Technological Innovations.ESA Working Paper No. 04-08 Agricultural and Development Economics Division The Food and Agriculture Organization of the United Nations 2004.

[18]   La Nacion. Campo La tecnología permite afirmar cada vez más la apertura en el calendario de siembra de maíz , [9th Dec 2015]; Available at: http://www.lanacion.com.ar/1852767- la-tecnologia-permi- e-afirmar-cada-vez-mas-la-apertura-en-el-calendario-de-siembra-de-maiz

[19]   Clarin. iEco Economía Diez años que revolucionaron la producción agrícola argentina , [1st Dec 2015]; Available at: http://www.ieco.clarin.com/economia/anos-revolucionaron-produccion-agricola-argentina_0_863313908.html

[20]   América economía. Economía & Mercados Producción agrícola crece 13% en Argentina en ciclo 2014-2015 , [28th Nov 2015]; Available at: http://www.americaeconomia.com/economia-mercados/finanzas/produccion-agricola-crece-13-en-argentina-en-ciclo-2014-2015

[21]   Altman A. Plant biotechnology in the 21st century: the challenges ahead. EJB Elect J Biotech 1999; 2(2): 51-5.

[22]   Ahloowalia BS, Prakash J, Savangikar VA, Savangikar C. Ahloowalia BS, Mohan Jain S Eds. Vienna: FAO/IAEA Division of Nuclear Techniques in Food and Agriculture 2004 Plant tissue culture, Proceedings of a Technical Meeting organized by the Joint FAO/IAEA Division of Nuclear Techniques in Food and Agriculture, Vienna, Austria 26–30 August 2002. 3-10.

[23]   Sussex IM. The scientific roots of modern plant biotechnology. Plant Cell 2008; 20(5): 1189-98. [http://dx.doi.org/10.1105/tpc.108.058735] [PMID: 18515500]

[24]   Chua NH, Sundaresan V. Plant biotechnology. The ins and outs of a new green revolution. Curr Opin Biotechnol 2000; 11(2): 117-9. [http://dx.doi.org/10.1016/S0958-1669(00)00069-0] [PMID: 10809543]

[25]   Akin-Idowu PE, Ibitoye DO, Ademoyegun OT. Tissue culture as a plant production technique for horticultural crops. Afr J Biotechnol 2009; 8(16): 3782-8.

[26]   Biofábrica Misiones S.A.. [28th Nov 2015]; Available at: http://www.biofabrica.com.ar

[27]   Centro biotecnológico Agrícola Forestal del Chaco. [28th Nov 2015]; Available at: http://www.escu eladejardineria.edu.ar/centro-bio-forestal.php

[28]   Debat HJ, Grabiele M, Aguilera PM, *et al.* Exploring the Genes of Yerba Mate (Ilex paraguariensis A. St.-Hil.) by NGS and De Novo Transcriptome Assembly PLoS ONE 2014; 9(10e109835): 1-16.

[29]   Censo Nacional Agropecuario 2008 - CNA'08. 2008 [1st Dec 2015]; Available at: http://www.indec.gov.ar/nuevaweb/cuadros/novedades/cna08_10_09.pdf

[30]   Censo Nacional de Población. Hogares y Viviendas , 2010 [1st Dec 2015]; Available at: http://www.indec.gov.ar/nivel4_default.asp?id_tema_1=2&id_tema_2=41&id_tema_3=135

[31]   Chiossone G. Sistemas de producción ganaderos del noreste argentino: Situación actual y propuestas tecnológicas para mejorar la productividad X Seminario de Pastos y Forrajes 2006 , 2006 [28th Nov 2015]; Available at: http://www.produccion-animal.com.ar/informacion_tecnica/origenes_evolucion_ y_estadisticas_de_la_ganaderia/65-Guillermo_Chiossone.pdf

[32]   Rau V. La yerba mate en misiones (Argentina): Estructura y significados de una producción localizada. Agroalimentaria 2009; 15(28): 49-58.

[33]   Bartolomé L. Colonos, plantadores y agroindustrias. Desarrollo Econ 1975; 15(58): 240-64.

[34]   Cáceres D. Los Sistemas Productivos de Pequeños Productores Tabacaleros y Orgánicos de la Provincia de Misiones. Estudios Regionales 2003; 23: 13-29.

[35]   Bolsi AS. Misiones: una aproximación geográfica al problema de la yerba mate y sus efectos en la ocupación del espacio y el poblamiento. Folia Histórica del Nordeste 1986; p. 7.

[36]   Gallero MC. Agroindustrias familiares en Misiones: Fábricas de ladrillo y almidón de mandioca de alemanes-brasileños Población y sociedad 2013; 20(1): 41-75.

[37]   Ahloowalia BS. Integration of technology from lab to land Proceedings of a Technical Meeting organized by the Joint FAO/IAEA Division of Nuclear Techniques in Food and Agriculture. Vienna, Austria. 2002; pp. 3-10.

[38]   Hussain A, Qarshi IA, Nazir H, Ullah I. Current Status and Opportunities. In: Leva A, Rinaldi LMR, Eds. Plant Tissue Culture. 2012. CC BY 3.0 license 2012. DOI: 10.5772/50568

[39]   Mroginski L, Sansberro P, Flaschland E. Biotecnología y Mejoramiento Vegetal II. In: Levitus G, Echenique V, Rubinstein C, Hopp E, Mroginski L, Eds. Establecimiento de cultivos de tejidos vegetales. Instituto Nacional de Tecnología Agropecuaria 2010; pp. 17-25.

CHAPTER 2

Gametic Embryogenesis, Somatic Embryogenesis, Plant Cell Cultures, and Protoplast Fusion: Progress and Opportunities in Biofuel Production

Marines Marli Gniech-Karasawa[*]

Università degli Studi di Palermo, Dipartimento di Scienze Agrarie e Forestali, Palermo, Italy

Abstract: Biofuel production represents an important alternative for replacing fossil fuels and reducing the emission of greenhouse gases into the atmosphere. With increasing demands for renewable fuel to replace fossil fuels, research on new energy sources is becoming more popular and new approaches in research techniques are occurring. In this context, the uses of somatic and gametic embryogenesis, cell suspension, and protoplast fusion in biofuels production will be presented in this chapter. Gametic embryogenesis is a convenient alternative in plant breeding because it makes possible the development of homozygous lines, increasing efficiency and speed in conventional breeding programs. Somatic embryogenesis is an important tool for plant cloning, looking toward the obtaining of improved plants by cell suspension culture or protoplast fusion. Suspension of plant cell cultures has several uses and applications for improving agronomical traits, and it is widely used in biotechnology for micropropagation, for the production of secondary metabolites or other substances, for obtaining somatic hybrids through protoplast fusion, and for modifying plants through genetic transformation. Protoplast fusion has been used by plant breeders to overcome the genetic barriers of outcrossing in incompatible plants, producing hybrid plants with different degrees of ploidy for improved agronomic and horticultural traits. In this chapter, current research with species that have potential to improve biofuel production is presented, with the aim of giving insights on the ways that these techniques can be used to produce renewable fuels.

Keywords: Bioenergy, Biotechnology, Cell suspension, Embryogenesis, Haploid technology, Somatic hybridization.

INTRODUCTION

Global climatic change and demographic pressure will continuously increase the demand for agronomic resources, and proving food and energy will therefore be

[*] **Corresponding author Marines Marli Gniech-Karasawa:** Università degli Studi di Palermo, Dipartimento di Scienze Agrarie e Forestali, Palermo, Italy; Tel: (+55)1698199-6800; E-mail: mgniechk@yahoo.com.br

Daniela Defavari do Nascimento, & William A. Pickering (Eds.)

one of the biggest challenges in plant production [1]. Along with this, the increasing consumption of fossil fuels and the reduction of resources are expected to increase oil prices. At the same time, the intensive and increasing use of fossil fuels has accelerated environmental degradation [2]. The most common plant species for producing biofuels in Europe and the United States is soybean; in Indonesia, rapeseed; in Malaysia and Thailand, palm oil; in Brazil, sugarcane is the most commonly used. All of them can compete with food production and bring about environmental problems [3].

In light of these problems [4], it is expected that microalgae will be the most promising alternative to replace agricultural crops producing sustainable biodiesel. However, according to Chen *et al.* [5], production costs are high and basic studies are necessary for elucidating microorganism characteristics and for developing microalgal biotechnology. In this context, C4 grasses from the Panicoideae clade must be included in the second generation production of bioethanol, for example: *Zea mays*, *Sorghum bicolor*, *Saccharum officinarum*, *Panicum virgatum*, and *Setaria viridis* [6, 7]. In addition, *Pennisetum purpureum*, *Arundo donax*, *Phalaris arundinacea*, *Mischantus* x *giganteus*, *M. sinensis*, *M. sacchariflorus*, *Eucaliptus globules*, *Jatropha curcas*, and *Pueraria Montana* species are listed as candidates for biofuel production, with research now in progress [8, 9]. Another sustainable pathway that has been exploited is the ability of *Mucor circillenoides* to convert single-cell oil into ethyl esters [10]. Another alternative is *Jatropha curcas* oil, which is less expensive for producing biodiesel [11, 12] and can be grown in poor environments [3]. *Populus deltoids* and *P. nigra* oils are also environmentally viable alternatives for reducing global warming, ozone depletion, and photochemical oxidation impact [13].

Biotechnology techniques present efficient alternatives for the large-scale production of clones for plant breeding and mutant selection. Using *in vitro* techniques, there are two alternatives for regenerating an entire plant from explants: organogenesis or somatic embryogenesis. Organogenesis can be obtained by inducing somatic explants to regenerate shoots and roots in appropriate culture conditions. On the other hand, somatic embryogenesis normally has a callus transition phase before embryo formation, and afterwards plant regeneration can be achieved on solid medium or by cell suspension in liquid medium. Cell suspension culture is an *in vitro* technique applied to isolated and multiple callus cells, where it is possible to produce large-scale plantlets or metabolic products of medicinal/industrial interest. For plant breeding, regarding the production of improved plants with different ploidy levels, it is possible to use the protoplast fusion technique. This procedure permits the production of hybrids for different purposes such as increased resistance, increased metabolism, seedless fruits, bigger fruits and flowers, *etc*. On other hand, if the breeding purpose is to obtain pure lines, selected mutants, and/or to produce diploid hybrids, the haploid technology can be used through gametic embryogenesis (Fig. **1**).

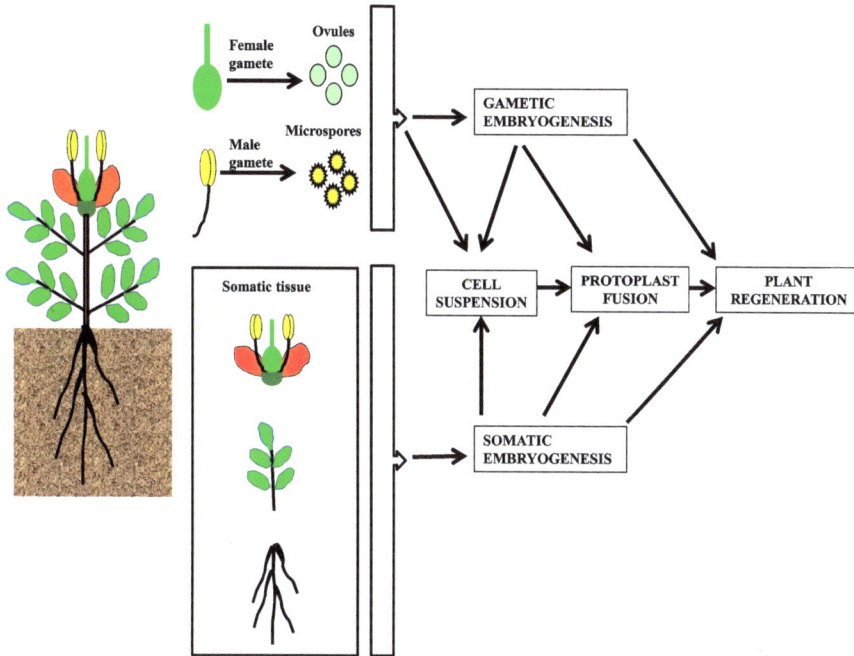

Fig. (1). Schematic representation showing a summary of the different ways of application of the techniques described in this chapter.

The purpose of this chapter is to expound the details of each technique and application as they relate to biofuel production. A hitherto unpublished procedure obtained with the oil producing plant *Olea europea* is presented in this article. Also presented is a protocol (see Appendix) for producing embryos from isolated gametes, with special attention given to the gametic embryogenesis (GE) technique.

GAMETIC EMBRYOGENESIS

Haploid technology, through gametic embryogenesis, is a promising and convenient alternative [14] that is being recognized as an important tool for plant breeding, making it possible to develop, in only one generation, completely homozygous plants from heterozygous parents [15]. The instantaneous production of homozygous plants through androgenesis (male gamete) or ginogenesis (female gamete) is highly appreciated as a practical perspective in research and plant breeding [16]. It makes it possible to shorten breeding cycles and fix agronomic traits [17] in pure lines, increasing the efficiency and speed of conventional breeding programs. It is also becoming an attractive system for transferring and obtaining stable integration of recombinant DNA in plant genomes, avoiding hemizygosis and saving time and resources in the production of transgenic plants [18]. Additionally, it is a useful tool for genetic selection and screening of recessive mutants [14], fast introgression of new agronomic traits, development of physical [19] and genetic maps [20 - 22], transformation and mutagenesis [23], reverse breeding [24, 25], genomic studies [26, 27], simplifying genome sequencing [28, 24, 25], and understanding totipotency and early fate decisions [29] in gamete cells.

The embryogenic process of male gametes has been widely studied. Mechanisms that coordinate the process of cell division and embryo formation are not well understood yet [29]. The most common procedure is anther culture, but haploid and doubled haploid plants still can be obtained through pistil culture and *in vitro* pollination, using irradiated pollen or pollen from triploid plants. To obtain haploid and doubled haploid plants, immature male and female gametes must be deviated from normal gametophytic to sporophytic development. For this, it is necessary to induce gametes through *in vitro* stress treatments, because this is the main factor in reprogramming development in cells and tissues toward the embryogenic pathway [30]. *In vitro* embryogenesis can be induced through different stresses, hot or cold temperature generally being the most utilized to induce embryo formation [31], and also considered the most effective [15].

The gametes subjected to stress will independently start the synthesis of heat shock proteins of small molecular weight (smHSPs), whether or not they have embryogenic competence, because this step is developmentally regulated. Other recent studies suggest the participation of histone lysine methyltransferase in dimethylation (H3K9me2) and in embryo cell differentiation and heterocromatinization events, while the histones H3Ac, H4Ac, and the histone acetyl transferase (HAT) should be involved in the activation of transcription, totipotency, and events related to the proliferation and reprogramming of the cell during embryo development [32]. In this regard, in order to have good embryogenic response in gamete induction it is necessary to isolate the correct nucleated stage [33]. After that, the microspores exposed to the inductive treatment will arrest or die [18]. According to Islam and Tuteja [34], efficient gametic embryogenesis is therefore normally induced by successful application of stress treatment.

The literature shows that plant regeneration has been achieved from gametic embryogenesis by submitting anthers/microspores of barley [35], rice, *Triticum* [36], *Lilium longiflorum* [37], *Citrus* [38 - 41], *Cucumis sativus* [42], *asparagus*, *Papaya* [43], *Malus* sp [44 - 47], and *Quercus* spp [48, 49] to low temperature. On the other hand, high temperature prompted good results for plant regeneration of *Brassica* [50, 51] and *Solanum* [52]. In olive, only pro-embryos were obtained at high temperature [53]. Recently, gametic embryos of hazelnut have been obtained from isolated microspore culture [54]. Regarding oil and bioenergy production crops, the most studied are *Brassica* species, with doubled haploid plants of *Brassica napus* [55] and *Linum usitatissimum* L. being efficiently regenerated [56]. For soybean, several researchers were able to obtain embryo-like structures and plantlet regeneration from anthers and isolated microspores [57]. Embryo-like structures have also been regenerated in other species currently being considered for bioenergy production, such as *Phleum pratense* [58], *Pennisetum typhoideum*, and *Olea europea*.

Induction Procedure for Gametic Embryogenesis

To induce gametic embryogenesis, it is necessary to collect material, identify the correct nucleated stage, and start pre-treatment of floral buds, depending on the species in use. After pre-treatment, the next steps comprise assepsy of flower buds, removal of petals and sepals, and isolation of anthers and/or pistils. If the interest is anther culture (Fig. **2**), anthers must be put in induction culture medium and subjected to stress treatment.

Fig. (2). Flower bud sterilization (left); anther culture (middle); flowers, freshly isolated anthers, and anthers with callus and embryo (right) in *Citrus* sp.

Alternatively, it is possible to pollinate the pistils by irradiated pollen or pollen from a triploid plant, for example, in order to induce ovule development without fertilization (Fig. **3**).

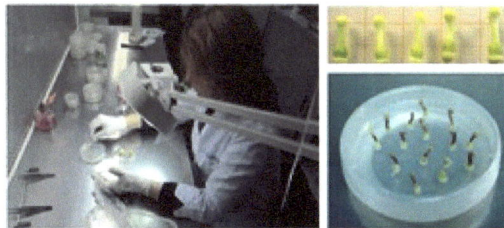

Fig. (3). *In vitro* pollination (left) and freshly isolated pistils and pistil culture (right) in *Citrus* sp.

If the purpose is to work with isolated gametes (microspores and/or ovules), it is necessary to remove the somatic tissues of anthers and pistils and perform a long and laborious method of work, until the gametes are isolated and purified (Fig. **4**). After that, the gametes need to be resuspended and the number of microspores counted, using a Burker chamber with an inverse microscope in order to adjust the density per milliliter (mL). Once this value is defined, we know how many petri dishes will be produced. The last step comprises subjecting them to stress treatment.

Fig. (4). Steps of isolated microspore culture in *Olea europea* sp.

From time to time, whatever gametic embryogenesis procedure has been chosen, (microspore/ovule), it is necessary to evaluate the response of gametes to the induction medium (Fig. **5**). This can be easily done through fluorescence microscopy, evaluating the nucleated stage and microspore behavior in the culture medium and evaluating the presence of calli and embryos by light microscopy (Fig. **6**).

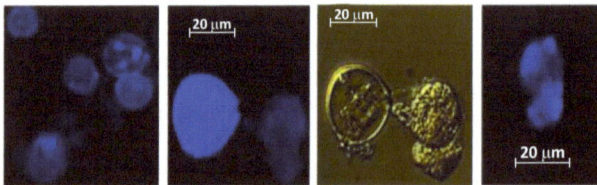

Fig. (5). Fluorescence microscopy study of isolated microspore showing the nucleated stage (left), microspore behavior in multinucleated stage by fluorescence and contrast of phases (middle), and presence of embryo (right).

Isolated Microspore Culture of *Olea Europea*

To induce microspore embryogenesis and understand developmental pathways in *Olea europea,* two different experiments were performed using P medium [59]. A protocol is provided in the Appendix to the present article. In the first experiment, the Meta-topolin (mT) experiment, six cultivars (cvs) (Galatina (Ga), Tonda Iblea (Ti), Bianchioline Pantenelleria (Bp), B. Napolitana (Bn), Oleastro (Ol), and Verdello (Vd)) were submitted to three stress temperatures (control (C), hot (H) 37°C/30', and freezing (F) -20°C/30') and five P medium supplemented with mT as a growth regulator (control (C), mT replacing benzilaminopurin (PmT/Ba), mT replacing zeatin (PmT/Zea), mT replacing benzilaminopurin ten times (PmT/Ba10), and mT replacing zeatin ten times (PmT/Zea10)). In the second experiment, the 2,4-D experiment, two cultivars (Galatina (Ga) and Tonda Iblea (Ti)) were submitted to the same stress temperatures described above, and to four P medium formulations with different 2,4-D concentrations (control (C), 1.0, 1.5, and 2.0 mg/L) for sixty days. After this period, they were replaced by P medium without 2,4-D. The mT experiment was delineated as a randomized block design, with six cultivars and fifteen treatments (three stress temperatures and five mT combinations). The 2,4-D experiment was also delineated as a randomized block design, with two cultivars and twelve treatments (three stress temperatures and four 2,4-D levels). All treatments were started with fifteen repetitions per treatment.

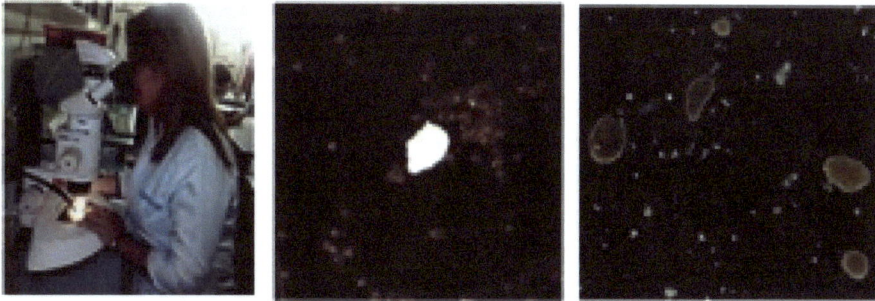

Fig. (6). Calli and embryo in isolated microspore culture in *Corylus avellana* sp.

Evaluations were done using a stereo-microscope (Leica MZ 125) and fluorescent microscopy, in order to observe the transition steps from gametophytic to sporophytic pathways.

During the induction phase, uninucleated, binucleated, trinucleated, and multinucleated cells were observed in microspore culture (Fig. **7**).

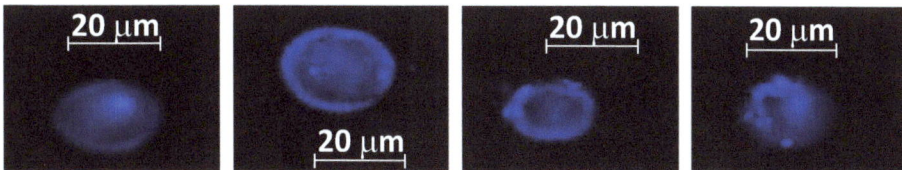

Fig. (7). Fluorescence microscopy study of isolated microspore showing nucleated stage in *Olea europea* sp.

After ten months of culture it was possible to observe the presence of pro-embryos, embryos, calli, and dead cells (Fig. **8**).

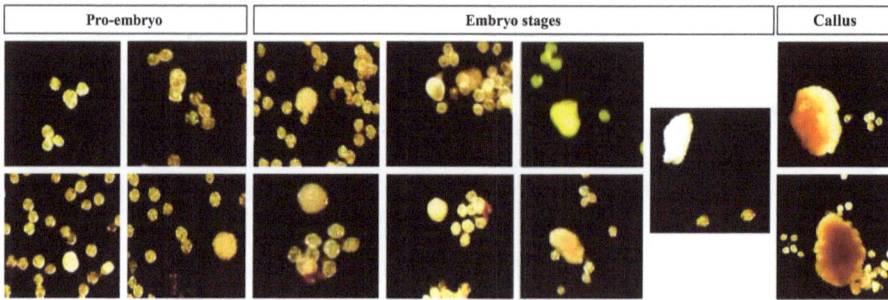

Fig. (8). Isolated microspore culture showing pro-embryo, embryo, and callus formation in *Olea europea* sp.

In the mT experiment, Tonda Iblea cultivar (Ti) had the better response to cell induction, resulting in more calli and embryos (ranging from 1 to 4), with a low number of dead cells at all mT levels (Fig. **9**). The highest efficiency induction/conversion calli/embryos were observed with hot stress treatment of Ti cultivar.

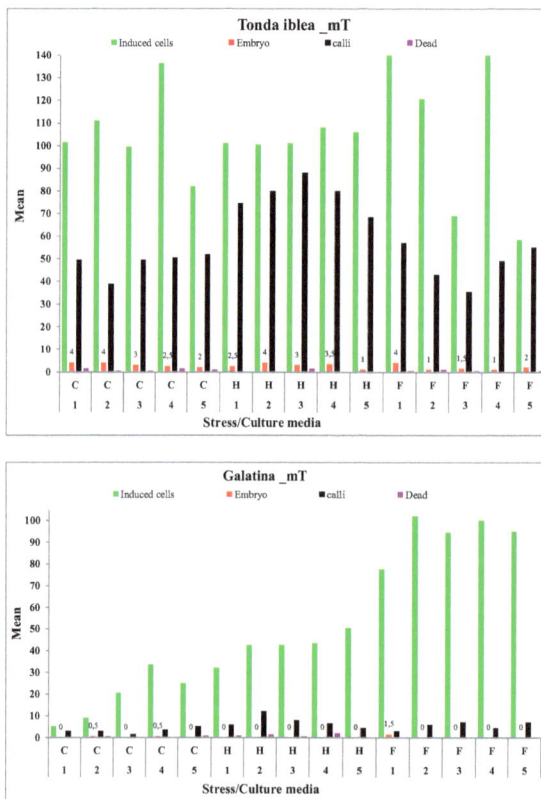

Fig. (9). Induced cells, calli and embryo production, and dead cells in Tonda Iblea (Ti) and Galatina (Ga) isolated microspore culture. 1: P medium, 2: PmT/Ba, 3: PmT/Zea, 4: PmT/Ba10, 5: PmT/ Zea10.

Galatina (Ga) was the least responsive cultivar in control temperature (C), but the best for cell induction at freezing stress (F) treatment. Ga cultivar presented the lowest number of calli in all treatments, with only a few embryos at control (C) temperature in P medium formulation supplemented with Meta-topolin replacing benzilminopurin (PmT/Ba and PmT/Ba10), and at P medium without regulators at freezing (F) temperature (Fig. **9**).

Bianchioline Pantenelleria (Bp) was the least responsive cultivar at hot (H) and freezing (F) stress temperatures, resulting in the highest number of dead cells and no embryos in any treatment (Fig. **10**).

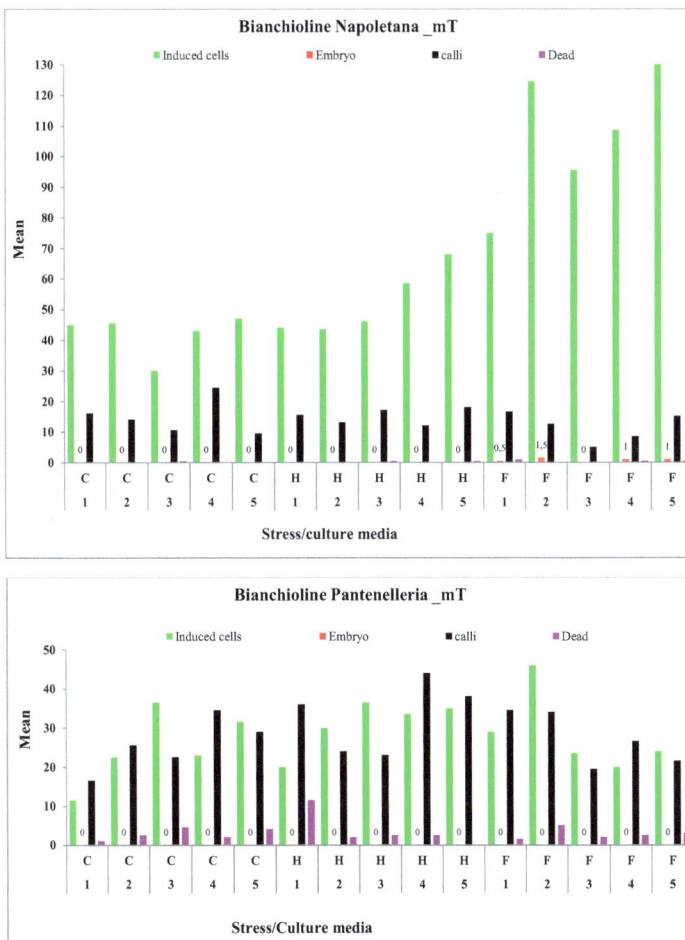

Fig. (10). Induced cells, calli and embryo production, and dead cells in Bianchioline Napoletana (Bn) and B. Pantenelleria (Bp) isolated microspore culture. 1: P medium, 2: PmT/Ba, 3: PmT/Zea, 4: PmT/Ba10, 5: PmT/Zea10.

High numbers of cell induction could be obtained with Bianchioline Napolitana (Bn) cultivar at freezing (F) treatment, but lower numbers of calli than Ti and Bp cultivars, with little embryo production at all freezing stress (F) treatments, except at PmT/Zea (Fig. **10**). It was possible to observe the conversion of induced cell to calli production in Bianchioline Pantenelleria (Bp) cultivar. However, no embryo was found, resulting in high numbers of dead cells (Fig. **10**). For the other two cultivars, Oleastro (Ol) and Verdello (Vd), no induced cells and no change were obtained during ten months in culture.

In the 2,4-D experiment, it was possible to observe a better conversion of calli and embryos from induced cells than in the mT (Meta-topolin) experiment. Tonda Iblea (Ti) was the cultivar in which the highest number of embryos was obtained (Fig. **11**), and the recorded number was higher than in the same cultivar at **mT** formulations (Fig. **9**). Until now, the greatest progress in obtaining gametic embryos has been achieved by Bueno *et al.* [53], who obtained only pro-embryos. To our knowledge, this is the first report of embryo formation using isolated microspore culture of olive.

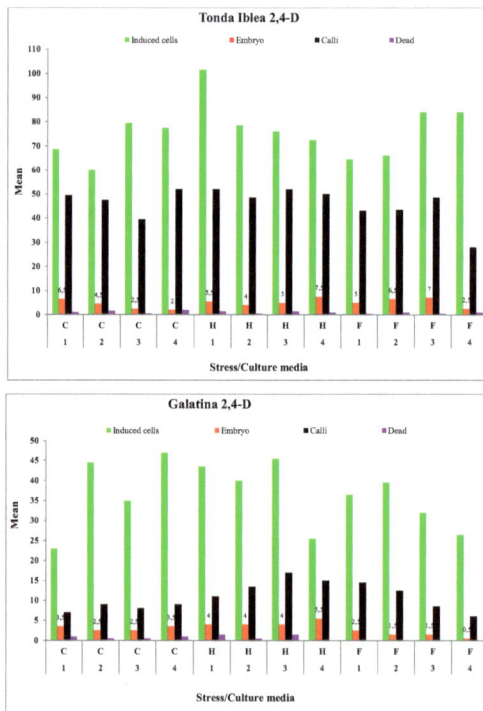

Fig. (11). Induced cells, calli and embryo production, and dead cells in Tonda Iblea (Ti) and Galatina (Ga) isolated microspore culture. 1: P medium, 2: P 2,4-D 1.0, 3: 2,4-D 1.5, 4: 2,4-D 2.0.

Comparing the mT and 2,4-D experiments, it is not possible to point out what is the best growth regulator concentration or the best stress treatment. But it seems that the best results for embryo production could be obtained for Ti cultivar under the following conditions: P medium at control temperature; 2.0 mg/L of 2,4-D with hot stress; and 1.5 and 2.0 mg/L of 2,4-D at freezing stress (Figs. **8**, **9**, and **11**). For Ga cultivar, control temperature and hot stress resulted in more embryos. Considering all results obtained, it seems that for Ga and Ti cultivars, 2,4-D and hot stress (37°C/30') is the best choice for gametic embryo production. For the Bn cultivar, use of Meta-topolin (mT) and freezing (F) treatment was able to produce few embryos. More studies are necessary for other cultivars (Bp, Bn, Ol, and Vd) in order to define a medium for producing embryos, and for Ga and Ti cultivars in order to optimize results.

SOMATIC EMBRYOGENESIS

Somatic embryogenesis is an *in vitro* morphogenic process in which autonomous embryos develop in response to exogenous and/or endogenous signals [60] through morphologic and chemical changes in a somatic cell [61] lacking the vascular connection with the original tissue [62, 63]. In theory, somatic embryogenesis can be started from any part of the donor plant, but young tissues and immature zygotic embryos show higher competence and produce better results. Somatic embryos can be originated from one cell or from a group of cells by plant tissue manipulation in *in vitro* conditions [64]. This technique can be used to reduce time, and offers an efficient system for clonal plant propagation, genetic transformation, and hybridization experiments, producing fewer chimeras [65]. Furthermore, in plant breeding it can be used together with traditional agriculture techniques [66].

Initiation of somatic embryogenesis comprises the induction of differentiated tissue to acquire embryogenic competence directly or indirectly [30]. Indirect somatic embryogenesis follows dedifferentiation of organized somatic tissue into a cell mass named callus, before embryo production. In direct somatic embryogenesis, the embryo will be produced directly from somatic tissue without the callus phase [67]. In a second step, competent cells will proliferate, producing pro-embryos and embryos (Fig. **12**). Then embryos will undergo the pre-maturation and maturation phases, and after that plant development will be achieved [62].

This technique has been used as a model procedure for understanding molecular events occurring during embryo development [68], because the developmental steps are highly similar to zygotic embryogenesis [69]. It is an important tool in the large-scale cloning of plants of high commercial value, and is considered a prerequisite for several biotechnological techniques during the production of improved plants [70], such as genetic transformation, somatic hybridization, and the achievement of somaclonal variation. According to Deo *et al.* [71], somatic embryogenesis and genetic transformation are indispensable tools in plant breeding, because they provide an alternative to developing control strategies against pests and diseases of agronomic crops.

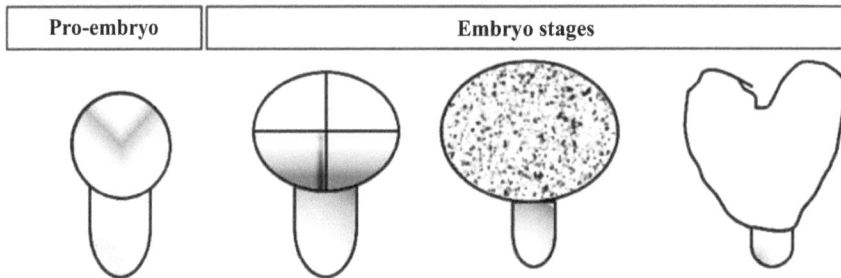

Fig. (12). Schematic representation of the somatic embryogenesis stages.

The regulation of somatic embryogenesis can be affected by several factors, such as hormones, proteins, transcription factors, stress treatments, plant species, health, age, *etc.* [30]. It is not yet well understood how plant hormones and growth regulators interact over somatic tissue, inducing the conversion to embryogenic tissue and allowing them to produce somatic embryos [72, 73]. However, the growth regulators 2,4-D, ABA, and ethylene are the most cited in affecting embryogenic potential [74, 75], and it has been shown that polyamines such as putrescine, spermine, and spermidine also affect embryogenic potential [76]. Considering the high number of inducers, it is widely accepted that somatic embryogenesis cannot be induced only by growth regulators, but that stress plays a critical role [30] in inducing somatic cells in the same way as occurs in gametic embryogenesis. According Fehér [60], all stress provoked *in vitro* will induce extensive and coordinated genetic reprogramming at the chromatin level, so that the developmental pathways will activate embryo formation through somatic embryogenesis. However, the way that stress treatments promote de-differentiation and produce competent cells that undergo embryogenesis is not known [68]. Furthermore, biochemical and molecular mechanisms that control the regulation of somatic embryo formation also remain poorly understood [63, 73]. For more details, see the review of Zavattieri *et al.* [30] that shows the effect of different stresses on the induction of somatic cells; see also Karami and Saidi [68], who discuss the molecular basis for stress acquisition in somatic embryogenesis, and Joshi and Kumar [73], who present the latest progress in regulation of somatic embryogenesis in crops.

Differential expression of genes producing mRNA and proteins was reported in the development of somatic embryos [73]. Also, genetic transformation using the genes WUSCHEL (WUS), BABY BOOM (BBM), Somatic Embryogenesis Receptor Kinase (SERK), and KNOTED-like homebox (KNOX), have been used to increase and induce somatic embryogenesis in recalcitrant plants [64]. PICKLE genes were described, acting on chromatin remodeling during the transition from embryogenic to vegetative development [77]. *Cyclin-dependent kinase* (CDKA) is expressed in embryogenic callus meristematic center [78]. LEC genes, originated from Arabidopsis leaf cotyledon, were reported to induce embryo development in vegetative cells.

Successful reports of somatic embryogenesis are found for different bioenergy and oil producing plants. With *Zea mays,* an efficient protocol was developed for *in vitro* regeneration of inbreed lines [74]; with *Saccharum officinarum* [79], for breeding and micropropagation purposes [80]; with *Pennisetum purpureum* [81], *Jatropha curcas* [82], and soybean [83], for establishing a culture medium for clonally micropropagating these species. According Ree and Guerra [84], protocols for somatic embryogenesis have been developed for economically important and endangered plants of the palm family; nineteen species are described using a wide range of explants and growth regulators.

PLANT CELL CULTURES

Conventional micropropagation techniques for plant cloning are time consuming and require intensive work. Therefore, in order to optimize plant tissue culture production, cell suspension culture has the purpose of helping in the acquirement and proliferation of cells [85], reducing time, and speeding regeneration of high-quality and commercially important material [86]. Since the end of the 1950s, plant cell suspension cultures have been increasingly used in basic research, and are currently widely applied in biotechnology to provide high-scale mass propagation, selection of mutants, and gene transfer [87, 65]. Proteomics studies, which identify proteins that can be changed by genetic engineering, are expected to increase the second generation of ethanol production [88]. Plant cell wall engineering is also expected to enhance biofuel production [89]. Examples of other areas with several uses and applications for improving agronomical traits, are protoplast fusion, somatic hybrids, and secondary metabolites. With respect to secondary metabolites, commercial attention generally is given to essential oils, glycosides, and alkaloids [90]. There is thus great interest in the design and optimization of plant cell cultures in order facilitate and accelerate the propagation of material with high agronomic value [91].

Cell suspension cultures are characterized by high cell concentration and aggregation, which lead to an increase in the viscosity of the medium [92]. These cultures are initiated by agitating cells in a liquid medium [93]. The cells can be isolated from embryogenic and non-embryogenic callus, or from isolated protoplasts. Embryogenic callus will regenerate plantlets *via* somatic embryogenesis, and non-embryogenic cells can be used for cell proliferation. After callus lines reach genetic stability, productivity evaluation can be done to select the best lines to be used in cell suspension [65]. Normally the whole process, from plant to stable cell suspension culture, takes from six to nine months, aggregates and cell clumps in the culture being usually achieved [94]. The ability of *in vitro* cultures to regenerate embryos is limited only to a group of cells, and the differential response among cell types can be assigned to time of exposure to growth regulators and the physiological state of each cell [63]. The advantage of this technique is that it can be continuously available, securing cell source supply [95].

The final step is large-scale propagation of bioreactors. The main use of bioreactors is for cell suspension cultures and secondary metabolite production. Recently, there has occurred an interest in the use of this system for clonal propagation [96] with industrial purposes. This is mainly because bioreactor systems permit the maintaining of large-scale plant cell cultures, enabling high quality and high yield of cell line production [91], reducing variations in the final product and simplifying the registration process. According to Yesil-Celiktas *et al.* [65], for bioreactors used for plant cell suspension cultures, the shaking system can be driven mechanically or pneumatically. Specific operating parameters can be altered in order to modulate the growth and function of cells, flow and mixing being the key factors responsible for the hydrodynamic effect on cell shape and function, because these factors change the transference of mass and nutrients [90]. To have optimal cell/plant propagation, it is necessary to better understand the biochemical and physiological responses of the culture environment, optimizing culture conditions [86]. Recently, modification in a disposable process has been designed to industrially propagate three million embryos per year of *Coffea canephora*. The main modification was the pre-germination modification of the embryos (previously done by temporary immersion in a 10L glass container) in one plastic bag bioreactor of 10L (box-i--bag) [97].

Among oil and biofuel producer species, Choi *et al.* [98], studied the effect of two operating parameters and observed that palm oil cell suspension had an increase of more than 200% in biomass production, when agitation rates ranged from 120-225 rpm and oxygenation was between 20% and 80%. Ibraheem *et al.* [99], studying different systems for propagating palm oil (*Phoenix dactylifera*), observed that a liquid system produced more somatic embryos in comparison with a temporary immersion and solid medium. Steinmacher *et al.* [100] obtained more embryo regeneration and secondary embryogenesis at temporary immersion than at solid medium. For microalgae, which have been reported as the most promising alternative in biofuel production [4], cell culture protocols are available, but according Perin *et al.* [101] these organisms should be genetically improved for increased photosynthetic activity and biomass production. Recently, Prancha *et al.* [102] increased the potential of biofuel production using *Scenedesmus* sp. by optimizing the carbon source and light intensity. However, the high costs of biodiesel production resulted in declining investment and research efforts regarding microalgae culture [5]. Another alternative strategy for increasing biodiesel production is the use of cellulase produced by submerged fermentation of fungi strains of *Trichoderma* sp. and *Lasiodiplodia theobromae* [105].

There are protocols available in the literature for cell suspension culture of many species (and with different purposes), such as *Jatropha curcas* [103] and 20 *J. gossypifolia*, whose biotechnological potential for bioactive compound production is being evaluated for pharmaceutical, agricultural, and industrial applications [64]. Cell suspension culture of *Helianthus annuus* was improved in bag bioreactors when compared to orbital shaken bioreactors, with less foam formation and less flotation [104]. Sugarcane embryonic cell suspension culture has already been established for calli derived from young leaves [79], and has already been optimized for stirring bioreactors [88].

PROTOPLAST FUSION

Plant protoplasts are cells whose cellular wall is removed by enzymes for *in vitro* manipulation [107]. Absence of cell wall is an interesting system for several experimental manipulations that are not possible with intact plant cells [108]. In theory, protoplasts can be isolated from any tissue, organ, or plant, but normally leaf tissues, calli, and cell suspension culture are used [109]. The absence of the cell wall makes protoplasts a useful system for genetic manipulation, and one that would not be possible with intact plants, in areas such as the following: production of haploid and doubled haploid plants, somatic hybrids, mutant selection, transgenesis, biochemical and molecular studies, *etc.*

Protoplast fusion has become an important tool for ploidy manipulation and for overcoming sexual barriers in conventional breeding [106].

To have high protoplast yield, osmotic potential (*i.e.,* sucrose, mannitol, sorbitol, glucose, galactose, *etc.*) must be optimized, as well as the proportion and type of enzyme (cellulase, hemicellulase, pectinase, pectoliase, *etc.*) in the solution [110]. Low enzyme concentration, low temperature, and high pH are normally used for a short time of incubation. Another important consideration is that protoplasts that are isolated in the presence of ions Ca^{++} and Mg^{++} have shown higher capacity of cell wall regeneration, as opposed to those isolated in the absence of these ions. Protoplast cells can be obtained from embryogenic callus or suspension culture. Cells from suspension culture provide better results if they are in the log growth phase [106]. The efficiency in removing cell wall is affected by plant genotype, the physiological state of cells, the type, concentration, and exposure time of enzymes, speed of shaking, and light and temperature of the environment [109, 111, 112].

After protoplast isolation, it is necessary to confirm viability. Once confirmed, it is possible to start the procedure steps for hybridization. Firstly, it is necessary to remove enzyme residues by filtering and centrifuging samples (Fig. **13**, middle). After centrifugation, the pellet remaining in the bottom of the tube is composed of viable and not viable protoplasts. This pellet will be diluted using a two phase osmotic solution that will bring the viable protoplasts to the middle phase region; not viable protoplasts will remain in the bottom of the tube.

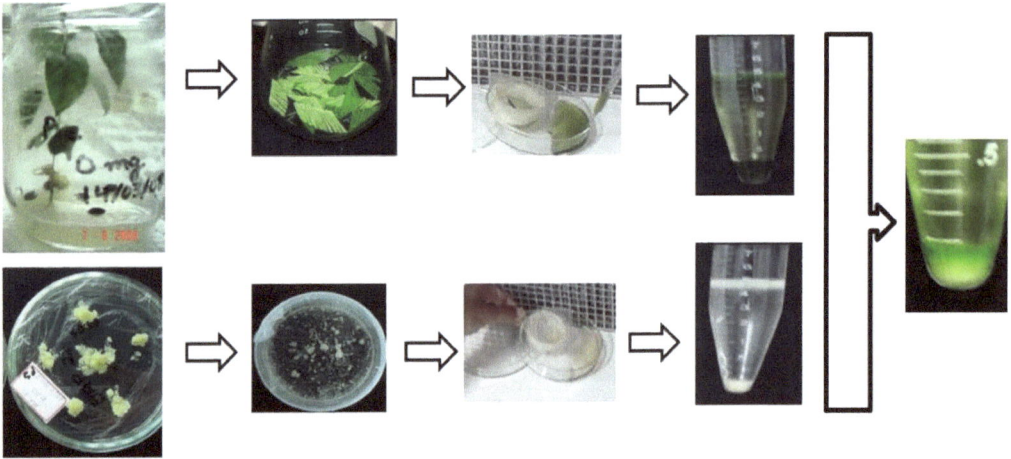

Fig. (13). Protoplast isolation (left) and hybridization (right) from leaves (above) and calli (below).

To produce hybrids, viable protoplast cells must be put together (Fig. **13**, right).

Three different techniques can be used to hybridize protoplasts: chemical fusion [113], electrical fusion [114], and electrochemical fusion [115]. All these techniques will promote fusion of the cells, using the instability of the plasmatic membrane [116]. The fusion is symmetric if the whole genome of both parents is incorporated during the process. It is asymmetric if only a part of the genome is included in the new individual [106].

For the hybridization procedure, protoplasts originating from leaves and from calli are normally used, because they allow us to confirm the fusion process and estimate the percentage of hybrid cells. After hybridization, unfused protoplasts, fused protoplasts with same genotype, and fused protoplasts of different genotypes will be obtained [107]. The progress of the culture will be observed by the formation of colony and cell wall deposition, and, after that, by microcally, embryo, and plant regeneration [117, 118].

Protoplast fusion provides an opportunity to avoid sexual barriers of sexual reproduction and permit taking nuclear and cytoplasm genomes to enrich the gene pool of cultivated species, generating a novel germplasm in conventional programs for elite breeding [119]. This technique has been used by plant breeders to overcome outcrossing genetic barriers in incompatible plants. It has been used with several approaches to produce plants with higher ploidy level and improved agronomic and horticultural traits, to create new cultivars, somatic hybrids, and transgenic plants, to select mutants, and for somaclonal variation [120].

Considering that agronomic traits such as oil content, yield, and disease resistance are governed by several genes (polygene), the production of transgenic plants using a single gene will not produce the desired results. In this sense, protoplast fusion remains a promising tool for transferring genes to cultivated species, for recovery of biotic and abiotic resistant plants, and for increased oil production and quality [119]. Protocols for producing protoplast cells and plant regeneration are available for a large number of vegetable, oil, and forage crops [121]. However, few successful reports are found using protoplast fusion to increase oil content. Among them, protoplast fusion between *B. napus* and *C. absinica* produced high erucic acid [119, 122], *B. napus* and *Camelina sativa* produced high linoleic acid [123], and *B. napus* and *Orychophragmus violaceus* improved oil quality [124].

Other reports using protoplast fusion to increase root resistance were found for the oil producer *Helianthus annuus* and its wild relative species *Helianthus maximiliani* and *Helianthus mollis* [125]. High yield of protoplast isolation and fusion was also achieved for *Pannicum virgatum* [87] and *Pennisetum purpureum* [126], with regard to increased cellulose production. Another study investigated optimal conditions for protoplast isolation in *Jatropha curcas* and *Ricinus communis,* in order to produce intergeneric hybrids [127], but no microcolony and plant regeneration were obtained [128]. Optimized somatic hybridization by electrofusion resulted in sugarcane microcallus after twenty days [129]. Liu *et al.* [130], have shown that is possible to obtain hybrid plants by fusion of haploid protoplasts obtained from microspore and from haploid mesophyll of *Brassica*. They claim that this technique can be useful for exploring genetic and cytological problems, for better understanding the effects of genome dosage, for reducing genomic ploidy level and increasing fertility, and for gene introgression between two species so as to produce new material for plant breeding.

APPENDIX
Protocol to work with isolated microspore culture of *Olea europea*

1. Collect flowers and submit them to 5°C for one week.

2. Sterilize axis and flowers with

a. Ethyl alcohol 70% for 3'.

b. Bleach solution (2.5% of active chlorine with a few drops of Tween 20) for 20'.

c. Rinse them three times in sterile distilled water.

3. Dissect the material carefully using forceps to remove about 500 flowers.

4. Put them in a petri dish containing ice cold mannitol 0.4M + 0.5g/L of ascorbic acid.

5. Transfer them to an ice cold blender and do two pulses of 20" each.

6. Filter the material with 100 mesh filters.

7. Transfer the material to a falcon tube of 50 mL and complete volume with mannitol.

8. Centrifuge at 7500 rpm for 10 minutes.

9. Remove the supernatant and add more mannitol (repeat this step 2x).

10. Resuspend the material in 20 mL of mannitol.

11. Take 20 µL and put it in the Burker chamber.

12. Carefully put a cover glass over the solution in the Burker chamber.

13. Count microspore number using inverse microscope.

14. Centrifuge microspores in falcon tube at 7500 rpm to remove mannitol.

15. Add **P** medium [59] supplemented with **2,4-D** or **mT** (replacing BA or ZEA), as described for *Olea europea* gametic embryo induction.

16. Proceed with the dilution, in order to obtain about 150,000 microspores/mL.

17. Distribute 1 mL of microspore solution in 30 mm petri dishes.

18. Seal petri dishes with parafilm and subject them to a stress temperature for 30', as described for *Olea europea* gametic embryos induction.

19. Leave them in the dark for 30 days and then transfer to light at 27 ±2°C.

20. After 60 days, remove liquid medium from the microspore culture if **2,4-D** was used, and add P medium without growth regulators.

21. Leave them in light at 27 ±2°C.

22. Add culture medium from time to time if necessary.

CONFLICT OF INTEREST

The author confirm that this chapter content has no conflict of interest.

ACKNOWLEDGEMENTS

Thanks to the Università degli Studi di Palermo for permitting the author's post-doctorate research and for providing plant material, reagents, laboratory, and supervisor to the author. In addition, thanks are due to the Conselho Nacional de Desenvolvimento Científico e Tecnológico (CNPq) for providing grant to the author (2013-2015) and the Fundação de Auxílio a Pesquisa do Estado da Bahia (FAPESB) for providing scholarships grant and financial support (2007-2009) to the author. Thanks also to Prof. Helaine Carrer for helping the author to make feasible the post-doctorate in Italy.

REFERENCES

[1]   Ochatt SJ. Agroecological impact of an *in vitro* biotechnology approach of embryo development and seed filling in legumes. Agron Sustain Dev 2015; 35: 535-52.
[http://dx.doi.org/10.1007/s13593-014-0276-8]

[2]   Atabani AE, Silitonga AS, Anjum BI. Mahlia TMI, Masjuki HH, Mekhilef S. A comprehensive review on biodiesel as an alternative energy resource and its characteristics. Renew Sustain Energy Rev 2012; 16(4): 2070-93.
[http://dx.doi.org/10.1016/j.rser.2012.01.003]

[3]   Silitonga AS, Atabani AE, Mahlia TM, Masjuki HH, Badruddin IA, Mekhilef S. A review on prospect of *Jatropha curcas* for biodiesel in Indonesi. Renew Sustain Energy Rev 2011; 15(8): 3733-56.
[http://dx.doi.org/10.1016/j.rser.2011.07.011]

[4]   Chisti Y. Biodiesel from microalgae beats bioethanol. Trends Biotechnol 2008; 26(3): 126-31.
[http://dx.doi.org/10.1016/j.tibtech.2007.12.002] [PMID: 18221809]

[5]   Chen H, Qiu T, Rong J, He C, Wang Q. Microalgal biofuel revisted: an informatics-based analysis of developments to date and future prospects. Appl Energy 2015; 155: 585-98.
[http://dx.doi.org/10.1016/j.apenergy.2015.06.055]

[6]   Petti C, Shearer A, Tateno M, *et al.* Comparative feedstock analysis in *Setaria viridis* L. as a model for C4 bioenergy grasses and Panicoid crop species. Front Plant Sci 2013; 4(181): 181.
[http://dx.doi.org/10.3389/fpls.2013.00181] [PMID: 23802002]

[7]   Weijde T, van der , Kamei CLA, Torres AF, *et al.* Front Plant Sci 2013; 4(107)
[http://dx.doi.org/doi: 10:3389/fpls.2013.00107]

[8]   Barney JN, DiTomaso JM. Global climate niche estimates for bioenergy crops and invasive species of agronomic origin: potential problems and opportunities. PLoS One 2011; 6(3): e17222.
[http://dx.doi.org/10.1371/journal.pone.0017222] [PMID: 21408056]

[9]   Singh MP, Erickson JP, Sollenberger SE, Woodard KR, Vendramini JM, Gilbert RA. Mineral composition and removal of six perennial grasses grown for bioenergy. Agron J 2015; 107(2): 466-74.
[http://dx.doi.org/10.2134/agronj14.0339]

[10]   Carvalho AK, Rivaldi JD, Barbosa JC, de Castro HF. Biosynthesis, characterization and enzymatic transesterification of single cell oil of Mucor circinelloidesa sustainable pathway for biofuel production. Bioresour Technol 2015; 181: 47-53.
[http://dx.doi.org/10.1016/j.biortech.2014.12.110] [PMID: 25625466]

[11]   Kumar A, Sharma S. An evaluation of multipurpose oil seed crop for industrial uses (*Jatropha curcas* L.): a review. Ind Crops Prod 2008; 28(1): 1-10.
[http://dx.doi.org/10.1016/j.indcrop.2008.01.001]

[12]   Berchmans HJ, Hirata S. Biodiesel production from crude *Jatropha curcas* L. seed oil with a high content of free fatty acids. Bioresour Technol 2008; 99(6): 1716-21.
[http://dx.doi.org/10.1016/j.biortech.2007.03.051] [PMID: 17531473]

[13]   Guo M, Li C, Facciotto G, *et al.* Bioethanol form poplar clone Imola: an environmentally viable alternative to fossil fuel? Biotechnol Biofuels 2015; 8(134)
[http://dx.doi.org/10.1186/s13068-015-0318-8]

[14]   Seguí-Simarro JM, Nuez F. How microspores transform into haploid embryos: changes associated with embryogenesis induction and microspore-derived embryogenesis. Physiol Plant 2008; 134(1): 1-12.
[http://dx.doi.org/10.1111/j.1399-3054.2008.01113.x] [PMID: 18507790]

[15]   Germanà MA. Gametic embryogenesis and haploid technology as valuable support to plant breeding. Plant Cell Rep 2011; 30(5): 839-57.
[http://dx.doi.org/10.1007/s00299-011-1061-7] [PMID: 21431908]

[16]   Germanà MA. Anther culture for haploid and doubled haploid production. Plant Cell Tissue Organ Cult 2011; 104: 283-300.
[http://dx.doi.org/10.1007/s11240-010-9852-z]

[17]   Data SK. Androgenic haploids: Factors controlling development and its application in crop improvement. Curr Sci 2005; 89(11): 1870-8.

[18]   Seguí-Simarro JM. Androgenesis revisited. Bot Rev 2010; 76: 377-404.
[http://dx.doi.org/10.1007/s12229-010-9056-6]

[19]   Van Leeuwen H, Monfort A, Zhang HB, Puigdomènech P. Identification and characterisation of a melon genomic region containing a resistance gene cluster from a constructed BAC library. Microcolinearity between *Cucumis melo* and *Arabidopsis thaliana*. Plant Mol Biol 2003; 51(5): 703-18.
[http://dx.doi.org/10.1023/A:1022573230486] [PMID: 12678558]

[20]   Houssain T, Tausend P, Graham G, Ho J. Registration of IBM2SYN10 doubled haploid mapping population of maize. J Plant Registrations 2007; 1: 81.
[http://dx.doi.org/10.3198/jpr2005.11.0414crs]

[21]   Chu CG, Xu SS, Friesen TL, Faris JD. Whole genome mapping in wheat doubled haploid population using SSRs and TRAPs and the identification of QTL for genomic traits. Mol Breed 2008; 22: 251-66. [http://dx.doi.org/10.1007/s11032-008-9171-9]

[22]   Zhang KP, Zhao L, Tian JC, Chen GF, Jiang XL, Liu B. A genetic map constructed using a doubled haploid population derived from two elite Chinese common wheat varieties. J Integr Plant Biol 2008; 50(8): 941-50. [http://dx.doi.org/10.1111/j.1744-7909.2008.00698.x] [PMID: 18713343]

[23]   Szareijo I, Foster BP. Doubled haploidy and induced mutation. Euphytica 2006; 158: 359-70. [http://dx.doi.org/10.1007/s10681-006-9241-1]

[24]   Dirks R, van Dun K, de Snoo CB, *et al.* Reverse breeding: a novel breeding approach based on engineered meiosis. Plant Biotechnol J 2009; 7(9): 837-45. [http://dx.doi.org/10.1111/j.1467-7652.2009.00450.x] [PMID: 19811618]

[26]   Talón M, Gmitter FG. Citrus genomics. Int J Plant Genomics 2008. Article n. 528361. [25] Ferrie AM, Möllers C. Haploids and doubled haploids in *Brassica* spp. for genetic and genomic research. Plant Cell Tissue Organ Cult 2011; 104: 375-86.

[27]   Forster BP, Heberle-Bors E, Kasha KJ, Touraev A. The resurgence of haploids in higher plants. Trends Plant Sci 2007; 12(8): 368-75. [http://dx.doi.org/10.1016/j.tplants.2007.06.007] [PMID: 17629539]

[28]   Aleza P, Juárez J, Hernández M, Pina JA, Ollitrault P, Navarro L. Recovery and characterization of a Citrus clementina Hort. ex Tan. Clemenules haploid plant selected to establish the reference whole Citrus genome sequence. BMC Plant Biol 2009; 9(110): 110. [http://dx.doi.org/10.1186/1471-2229-9-110] [PMID: 19698121]

[29]   Soriano M, Li H, Boutilier K. Microspore embryogenesis: establishment of embryo identity and pattern in culture. Plant Reprod 2013; 26(3): 181-96. [http://dx.doi.org/10.1007/s00497-013-0226-7] [PMID: 23852380]

[30]   Zavattieri MA, Frederico AM, Lima M, Sabino R, Arnhold-Schmitt B. Induction of somatic embryogenesis as an example of stress-related plant. Electron J Biotechnol 2010; 13(1): 1-9. [http://dx.doi.org/10.2225/vol13-issue1-fulltext-4]

[31]   Ferrie AM, Caswell KL. Isolated microspore culture techniques and recent progress for haploid and doubled haploid plant production. Plant Cell Tissue Organ Cult 2010; 104(3): 301-9. [http://dx.doi.org/10.1007/s11240-010-9800-y]

[32]   Rodríguez-Sanz H, Moreno-Romero J, Solís MT, Köhler C, Risueño MC, Testillano PS. Changes in histone methylation and acetylation during microspore reprogramming to embryogenesis occur concomitantly with Bn HKMT and Bn HAT expression and are associated with cell totipotency, proliferation, and differentiation in *Brassica napus*. Cytogenet Genome Res 2014; 143(1-3): 209-18. [http://dx.doi.org/10.1159/000365261] [PMID: 25060767]

[33]   Pechan PM, Smykal P. Androgenesis: affecting the fate of the male gametophyte. Physiol Plant 2001; 111: 1-8. [http://dx.doi.org/10.1034/j.1399-3054.2001.1110101.x]

[34]   Islam SM, Tuteja N. Enhancement of androgenesis by abiotic stress and other pretreatments in major crop species. Plant Sci 2012; 182: 134-44. [http://dx.doi.org/10.1016/j.plantsci.2011.10.001] [PMID: 22118624]

[35]   Hou L, Ullrich SE, Kleinhofs A, Stiff CM. Improvement of anther culture methods for doubled haploid production in barley breeding. Plant Cell Rep 1993; 12(6): 334-8. [http://dx.doi.org/10.1007/BF00237430] [PMID: 24197259]

[35]   Lentini Z, Reys P, Martínes CP, Nuñez VM, Roca WM. Mejoramiento del arroz com cultivo de anteras: applicaciones en el desarollo de germoplasma adaptado a ecosistemas Latinoamericanos y el Caribe. Cali: CIAT 1994; p. 79.

[36]   Labbani Z, Richard N, De Buyser J, Picard E. Plantes chlorophylliennes de blé dur obtenues par culture de microspores isolées : importance des prétraitements. C R Biol 2005; 328(8): 713-23.
[http://dx.doi.org/10.1016/j.crvi.2005.05.009] [PMID: 16125649]

[37]   Arzate-Fernandez AM, Nakazaki T, Yamagata H, Tanisaka T. Production of doubled-haploid plants from *Lilium longiflorum* Thunb. anther culture. Plant Sci 1997; 123: 179-87.
[http://dx.doi.org/10.1016/S0168-9452(96)04573-6]

[38]   Chen Z. A study on induction of plants from *Citrus* pollen. Fruit Varieties Journal 1985; 39: 44-50.

[39]   Geraci G, Starrantino A. Attempt to regenerate haploid plants from *in vitro* cultures of Citrus anthers. Acta Hortic 1990; 280: 315-20.
[http://dx.doi.org/10.17660/ActaHortic.1990.280.53]

[40]   Froelicher Y, Ollitrault P. Effects of the hormonal balance on *Clausena excavata* androgenesis. Acta Hortic 2000; 535: 139-46.
[http://dx.doi.org/10.17660/ActaHortic.2000.535.16]

[41]   Germanà MA, Chiancone B. Improvement of *Citrus clementina* Hort. ex Tan. microspore-derived embryoid induction and regeneration. Plant Cell Rep 2003; 22(3): 181-7.
[http://dx.doi.org/10.1007/s00299-003-0669-7] [PMID: 12879259]

[42]   Ashok Kumar HG, Murthy HN, Paek KY. Embryogenesis and plant regeneration from anther cultures of *Cucumis sativus* L. Sci Hortic (Amsterdam) 2003; 98: 213-22.
[http://dx.doi.org/10.1016/S0304-4238(03)00003-7]

[43]   Rimberia FK, Sunagawa H, Urasaki N, Ishimine Y, Adaniya S. Embryo induction *via* anther culture in papaya and sex analysis of the derived plantlets. Sci Horticult 2005; 103: 199-208.
[http://dx.doi.org/10.1016/j.scienta.2004.04.013]

[44]   Höfer M, Touraev A, Heberle-Bors E. Induction of embryogenesis for isolated microspores. Plant Cell Rep 1999; 18: 1012-7.
[http://dx.doi.org/10.1007/s002990050700]

[45]   Höfer M. *In vitro* androgenesis in apple In: Malusymski M, Ed. Doubled Haploid Production in Crop Plants. Kluwer academic publishers. IAEA 2003; pp. 287-92.

[46]   Höfer M. *In vitro* androgenesis in apple improvement of the induction phase. Plant Cell Rep 2004; 22(6): 365-70.
[http://dx.doi.org/10.1007/s00299-003-0701-y] [PMID: 14685764]

[47]   Zhang C, Tsukuni T, Ikeda M, *et al.* Effects of microspore developmental stage and cold pre-treatment of flower buds on embryo induction in Apple (*Malus* x *domestica Borkh.*) anther culture. J Jpn Soc Hortic Sci 2013; 82(2): 114-24.
[http://dx.doi.org/10.2503/jjshs1.82.114]

[48]   Bueno MA, Agundez MD, Gomez A, Carrascosa MJ, Manzanera JA. Haploid origin of cork-oak anther embryos detected by enzyme and rapid gene markers. Int J Plant Sci 2000; 161(3): 363-7.
[http://dx.doi.org/10.1086/314265] [PMID: 10817971]

[49]   Bueno MA, Gómez A, Sepúlveda F, *et al.* Microspore-derived embryos from *Quercus suber* anthers mimic zygotic embryos and maintain haploidy in long-term anther culture. J Plant Physiol 2003; 160(8): 953-60.
[http://dx.doi.org/10.1078/0176-1617-00800] [PMID: 12964871]

[50]   Dunwell JM, Thurling N. Role of sucrose in microspore embryo production in *Brassica napus* spp. *oleifera*. J Exp Bot 1985; 36: 1478-91.
[http://dx.doi.org/10.1093/jxb/36.9.1478]

[51]   Burnett L, Yarrow S, Huang B. Embryogenesis and plant regeneration from isolated microspores of *Brassica rapa* L. ssp. Oleifera. Plant Cell Rep 1992; 11(4): 215-8.
[http://dx.doi.org/10.1007/BF00232537] [PMID: 24202990]

[52]   Cappadocia M, Cheng DS, Ludlum-Simonette R. Plant regeneration from *in vitro* culture of anthers of Solanum chacoense Bitt. and interspecific diploid hybrids S. *tuberosum* L. x S. *chacoense* Bitt. Theor Appl Genet 1984; 69(2): 139-43.
       [PMID: 24253704]

[53]   Bueno MA, Pintos B, Martin A. Induction of embryogenesis *via* isolated microspore culture in *Olea europaea* L Olivebioteq 2006. Mazara del Vallo, Marsala, Italy

[54]   Gniech Karasawa MM, Chiancone B, Gianguzzi V, Albelgalel AM, Germanà MA. Gametic embryogenesis through isolated microspore in *Corylus avellana* L. Plant Cell Tissue Organ Cult 2015; 124: 1-13.

[55]   Prem D, Solís MT, Bárány I, Rodríguez-Sanz H, Risueño MC, Testillano PS. A new microspore embryogenesis system under low temperature which mimics zygotic embryogenesis initials, expresses auxin and efficiently regenerates doubled-haploid plants in *Brassica napus*. BMC Plant Biol 2012; 12: 127.
       [http://dx.doi.org/10.1186/1471-2229-12-127] [PMID: 22857779]

[56]   Obert B, Zácková Z, Samaj J, Pretová A. Doubled haploid production in Flax (*Linum usitatissimum* L.). Biotechnol Adv 2009; 27(4): 371-5.
       [http://dx.doi.org/10.1016/j.biotechadv.2009.02.004] [PMID: 19233256]

[57]   Crosser JS, Lülsdorf MM, Davies PA, *et al.* Toward doubled haploid production in the fabaceae: progress, constraints, and opportunities. Crit Rev Plant Sci 2007; 25: 139-57.
       [http://dx.doi.org/10.1080/07352680600563850]

[58]   Abdullah AA, Pedersen S, Andersen SB. Triploids and hexaploid regenerants from hexaploid thimoty (*Phleum pretense* L.) *via* anther culture. Plant Breed 1994; 112: 342-5.
       [http://dx.doi.org/10.1111/j.1439-0523.1994.tb00694.x]

[59]   Germanà MA, Scarano MT, Crescimanno FG. First results on isolated microspore culture of Citrus. Proc Int Soc Citricult 1996; 2: 882-5.

[60]   Fehér A. Somatic embryogenesis – Stress-induced remodeling of plant cell fate. Biochimica et Biophysica Acta 1840; 385-402.

[61]   Nawrot-Chorabik K. Somatic embryogenesis in forest plants In: Sato K-I, Ed. Embriogenesis: Tech: Open Science 20, 2012. Available at: http://www.Intechopen.com/books/embryogenesis//somatic-embryogenesis-in-woody-plants

[62]   von Arnold S, Sabala I, Bozhkov P, Dyachok J, Filanova L. Developmental pathways in somatic embryogenesis. Plant Cell Tissue Organ Cult 2002; 69: 233-49.
       [http://dx.doi.org/10.1023/A:1015673200621]

[63]   Quiroz-Figueroa FR, Rojas-Herrera R, Galaz-Avalos RM, Loyola-Vargas VM. Embryo production through somatic embryogenesis can be used to study cell differentiation in plants. Plant Cell Tissue Organ Cult 2006; 86: 285-301.
       [http://dx.doi.org/10.1007/s11240-006-9139-6]

[64]   Sólis-Ramos LY, Andrade-Torres A, Carbonell LA, Opereza Salín CM, de la Serna EC. Somatic embryogenesis in recalcitrant plants. In: Sato K-I, Ed. EmbriogenesisTech: Open Science 2012.25 Available at: http://www.Intechopen.com/books/embryogenesis//somatic-embryogenesis-in-woody-plants

[65]   Yesil-Celiktas O, Gurel A, Vardar-Sukan F. Large scale cultivation of plant cell and tissue culture in bioreactors. Transworld Research Network 2010; 37/661(2): 1-54.

[66]   De-la-Peña C, Nic-Can GI, Galaz-Ávalos RM, Avilez-Montalvo R, Loyola-Vargas VM. The role of chromatin modifications in somatic embryogenesis in plants. Front Plant Sci 2015; 6: 635.
       [http://dx.doi.org/10.3389/fpls.2015.00635] [PMID: 26347757]

[67]   Slater A, Scott NW, Fowler MR. Plant Biotechnology. England: Oxford University Press 2003.

[68]   Karami O, Saidi A. The molecular basis for stress-induced acquisition of somatic embryogenesis. Mol Biol Rep 2010; 37(5): 2493-507.
[http://dx.doi.org/10.1007/s11033-009-9764-3] [PMID: 19705297]

[69]   Zimmerman JL. Somatic embryogenesis: a model for early development in higher plants. Plant Cell 1993; 5(10): 1411-23.
[http://dx.doi.org/10.1105/tpc.5.10.1411] [PMID: 12271037]

[70]   Gutiérrez-Mora A, González-Gutiérrez AG, Rodríguez-Garay B, Ascencio-Cabral A, Li-Wei L. Plant somatic embryogenesis: some useful considerations. In: Sato K-I, Ed. EmbriogenesisTech: Open Science 2012.10 Available at: http://www.Intechopen.com/books/embryogenesis//somatic-embryogenesis-in-woody-plants
[http://dx.doi.org/10.5772/36345]

[71]   Deo PC, Tyagi AP, Taylor M, Harding R, Becker D. Factors affecting somatic embryogenesis and transformation in modern plant breeding. South Pacific J Nat Appl Sci 2010; 28: 27-40.
[http://dx.doi.org/10.1071/SP10002]

[72]   Jiménes VM. Involvement of plant hormones and plant growth regulators on *in vitro* somatic embryogenesis. Plant Growth Regul 2005; 47: 91-110.
[http://dx.doi.org/10.1007/s10725-005-3478-x]

[73]   Joshi R, Kumar P. Regulation in crops somatic embryogenesis: a review. Agricult Rev 2013; 34(1): 1-20.

[74]   Anami SE, Mgutu AJ, Taracha C, *et al.* Plant Cell Tissue Organ Cult 2010; 102: 285-95.
[http://dx.doi.org/10.1007/s11240-010-9731-7]

[75]   Baskaran P, Staden JV. Somatic embryogenesis of *Mervilla plumbea* (Lindl.) Speta. Plant Cell Tissue Organ Cult 2012; 109: 517-24.
[http://dx.doi.org/10.1007/s11240-012-0118-9]

[76]   Wu XB, Wang J, Liu J-H, Deng XX. Involvement of polyamine biosynthesis in somatic embryogenesis of Valencia sweet orange (*Citrus sinensis*) induced by glycerol. J Plant Physiol 2009; 166(1): 52-62.
[http://dx.doi.org/10.1016/j.jplph.2008.02.005] [PMID: 18448195]

[77]   Ogas J, Kaufmann S, Henderson J, Somerville C. PICKLE is a CHD3 chromatin-remodeling factor that regulates the transition from embryonic to vegetative development in *Arabidopsis.* Proc Natl Acad Sci USA 1999; 96(24): 13839-44.
[http://dx.doi.org/10.1073/pnas.96.24.13839] [PMID: 10570159]

[78]   Montero-Cortés M, Rodríguez-Paredes F, Burgueff C, *et al.* Characterization of a cyclin-dependent kinase (CDKA) gene expressed during somatic embryogenesis of coconut palm. Plant Cell Tissue Organ Cult 2010; 102: 251-8.
[http://dx.doi.org/10.1007/s11240-010-9714-8]

[79]   Ho WJ, Vasil IK. Somatic embryogenesis in sugarcane (*Saccharum officinarum* L.): growth and plant regeneration from embryogenic cell suspension culture. Ann Bot (Lond) 1982; 51: 719-26.
[http://dx.doi.org/10.1093/oxfordjournals.aob.a086523]

[80]   Reis RS, Vale E de M, Heringer AS, Santa-Catarina C, Silveira V. Putrescine induces somatic embryo development and proteomic changes in callus of sugarcane. J Proteomics 2015; 130: 170-9.
[http://dx.doi.org/10.1016/j.jprot.2015.09.029] [PMID: 26435420]

[81]   Haydu Z, Vasil IK. Somatic embryogenesis and plant regeneration from leaf tissues and anthers of *Pennisetum purpureum Schum.* Theor Appl Genet 1981; 59(5): 269-73.
[http://dx.doi.org/10.1007/BF00264978] [PMID: 24276510]

[82]   Feitosa LS, da Costa AS, Arrigoni-Blank M de F, Dibax R, Botanico MP, Blank AF. Indução e análise histológica de calos em explantes foliares de *Jatropha curcas* L. (Euphorbiaceae). Biosci J 2013; 29(2): 370-7.

[83]   Thibaud-Nissen F, Shealy RT, Khanna A, Vodkin LO. Clustering of microarray data reveals transcript patterns associated with somatic embryogenesis in soybean. Plant Physiol 2003; 132(1): 118-36.
        [http://dx.doi.org/10.1104/pp.103.019968] [PMID: 12746518]

[84]   Ree JF, Guerra MP. Palm (Arecaceae) somatic embryogenesis. *In Vitro* Cell Dev Biol Plant 2015.
        [http://dx.doi.org/10.1007/s11627-015-9722-9]

[85]   Cid LP. Suspensão celular. In: Torres AC, Caldas LS, Buso JA, Eds. Cultura de Tecidos e transformação Genética de Plantas. Brasília: Embrapa-SPI/Embrapa-CNPH 1998.

[86]   Mehrotra S, Goel MK, Kukreja AK, Mishra BN. Efficiency of liquid culture systems over conventional micropropagation: a progress towards commercialization. Afr J Biotechnol 2007; 6(13): 1484-92.

[87]   Mazarei M, Al-Ahmad H, Rudis MR, Joyce BL, Stewart CN Jr. Switchgrass (*Panicum virgatum* L.) cell suspension cultures: Establishment, characterization, and application. Plant Sci 2011; 181(6): 712-5.
        [http://dx.doi.org/10.1016/j.plantsci.2010.12.010] [PMID: 21958714]

[88]   Calderan-Rodrigues MJ, Jamet E, Bonassi MB, *et al.* Cell wall proteomics of sugarcane cell suspension cultures. Proteomics 2014; 14(6): 738-49.
        [http://dx.doi.org/10.1002/pmic.201300132] [PMID: 24436144]

[89]   Loqué D, Scheller HV, Pauly M. Engineering of plant cell walls for enhanced biofuel production. Curr Opin Plant Biol 2015; 25: 151-61.
        [http://dx.doi.org/10.1016/j.pbi.2015.05.018] [PMID: 26051036]

[90]   Sajc L, Grubisic D, Vunjak-Novakovic G. Bioreactors for plant engineering: an outlook for the future research. Biochem Eng J 2000; 4: 89-99.
        [http://dx.doi.org/10.1016/S1369-703X(99)00035-2]

[91]   Khalid N, Chee WW, Bokhari F, *et al.* Banana, gingers and papaya cell cultures for high throughput agriculture. Asia Pac J Mol Biol Biotechnol 2010; 18(1): 73-4.

[92]   Kieran PM, MacLoughlin PF, Malone DM. Plant cell suspension cultures: some engineering considerations. J Biotechnol 1997; 59(1-2): 39-52.
        [http://dx.doi.org/10.1016/S0168-1656(97)00163-6] [PMID: 9487717]

[93]   Wang Z, Hopkins A, Mian R. Forage and turf grass biotechnology. Crit Rev Plant Sci 2001; 20(6): 573-619.
        [http://dx.doi.org/10.1016/S0735-2689(01)80005-8]

[94]   Mustafa NR, de Winter W, van Iren F, Verpoorte R. Initiation, growth and cryopreservation of plant cell suspension cultures. Nat Protoc 2011; 6(6): 715-42.
        [http://dx.doi.org/10.1038/nprot.2010.144] [PMID: 21637194]

[95]   Subhashini P, Raja S, Thangaradjou T. Establishment of cell suspension culture protocol for seagrass (*Halodule pinifolia*): Growth kinetics and hidromorphological characterization. Aquat Bot 2014; 117: 33-40.
        [http://dx.doi.org/10.1016/j.aquabot.2014.04.005]

[96]   Gurel A. Propagation possibilities of commercially important plants through tissue culture techniques. Int J Natural Eng Sci 2009; 3(2): 7-9.

[97]   Ducos JP, Terrier B, Courtois D. Disposable bioreactors for plant micropropagation and mass plant cell culture. Adv Biochem Eng Biotechnol 2009; 115: 89-115.
        [http://dx.doi.org/10.1007/10_2008_28] [PMID: 19475375]

[98]   Choi DS, Andrade MHC, Willis LB, *et al.* Effect of agitation and aeration on yield optimization of palm oil suspension culture. Journal of Oil Palm Research Special Issue on Malasia-MIT Biotechnology Partnership programme 2008; 1: 23-34.

[99] Ibraheem Y, Pinker I, Böhme M. A comparative study between solid and liquid medium culture relative to callus growth and somatic embryo formation in date palm (*Phoenix dactylifera* L.) cv. Zaghlool. Emirate J Food and Agricult 2013; 25: 883-98.
[http://dx.doi.org/10.9755/ejfa.v25i11.16661]

[100] Steinmacher DA, Guerra MP, Saare-Surminski K, Lieberei R. A temporary immersion system improves *in vitro* regeneration of peach palm through secondary somatic embryogenesis. Ann Bot (Lond) 2011; 108(8): 1463-75.
[http://dx.doi.org/10.1093/aob/mcr033] [PMID: 21355009]

[101] Perin G, Bellan A, Segalla A, Meneghesso A, Alboresi A, Morosinotto T. Generation of random mutants to improve light-use efficiency of *Nannochloropsis gaditana* cultures for biofuel production. Biotechnol Biofuels 2015; 8(161): 161.
[http://dx.doi.org/10.1186/s13068-015-0337-5] [PMID: 26413160]

[102] Pancha I, Chokshi K, Mishra S. Enhanced biofuel production potential with nutritional stress amelioration through optimization of carbon source and light intensity in *Scenedesmus* sp. CCNM 1077. Bioresour Technol 2015; 179: 565-72.
[http://dx.doi.org/10.1016/j.biortech.2014.12.079] [PMID: 25579231]

[103] Wang Q, Cheng J, Zhan P, Zhan L, Kong Q. Establishment of suspension cell system for transformation of *Jatropha curcas* using nanoparticles. Adv Mat Res 2013; 608-609: 314-9.
[http://dx.doi.org/10.4028/www.scientific.net/AMR.608-609.314]

[104] Werner S, Greulich J, Geipel K, Steingroewer J, Bley T, Eibl D. Mass propagation of *Helianthus annuus* of suspension cell in orbitally shaken bioreactors: improved growth rate in single-use bag bioreactors. Eng Life Sci 2014; 14: 676-84.
[http://dx.doi.org/10.1002/elsc.201400024]

[105] Faheina Junior G da S. Amorin MV da FS, Sousa CG de, Sousa KA de, Saavedra GA. Strategies to increase cellulase production using fungi isolated from the Brazilian biome. Acta Scientarum Biological Science 2015; 37(1): 15-22.
[http://dx.doi.org/10.4025/actascibiolsci.v37i1.23483]

[106] Grosser JW, Gmitter FG Jr. Protoplast fusion for production of tetraploids and triploids: applications for scion and rootstock breeding in Citrus. Plant Cell Tissue Organ Cult 2011; 104: 343-57.
[http://dx.doi.org/10.1007/s11240-010-9823-4]

[107] Carneiro VT de C. Conroi T, Barros LMG, Matsumoto K Protoplastos: cultura e aplicações. Brasília: Embrapa-SPI/Embrapa-CNPH 1998.

[108] Fungaro MH, Vieira ML. Protoplastos de plantas: isolamento e regeneração. Cienc Cult 1989; 41: 1151-9.

[109] Szabados L. Protoplastos: aislamiento, cultivo y regeneración de plantas. In: Rocca WM, Mroginski LA, Eds. Cultivo de tejidos en la agricultura: fundamentos y aplicaciones. Cali: CIAT 1993; pp. 239-70.

[110] Tomar UK, Dantu P. Protoplast fusion and somatic hybridization. Cellular and Biochemical Science Book. 2009; pp. 41876-91.

[111] Evans DA, Bravo JE. Protoplast isolation and culture. In: Evans DA, Sharp WR, Ammirato PV, Yamada Y, Eds. Handbook of plant cell culture. New York: McMillan Press 1983; Vol. 1: pp. 124-76.

[112] Gleddie SC. Protoplasts and transformation procedures. In: Gamborg OL, Phillips GC, Eds. Plant cell tissue and organ culture: fundamental methods. Berlin: Springer-Verlag 1995; pp. 167-80.
[http://dx.doi.org/10.1007/978-3-642-79048-5_14]

[113] Kao KN, Michayluk MR. A method for high-frequency intergeneric fusion of plant protoplasts. Planta 1974; 115(4): 355-67.
[http://dx.doi.org/10.1007/BF00388618] [PMID: 24458930]

[114] Zimmermann U, Scheurich P. High frequency fusion of plant protoplasts by electric fields. Planta 1981; 151(1): 26-32.
[http://dx.doi.org/10.1007/BF00384233] [PMID: 24301666]

[115] Olivares-Fuster O, Duran-Vila N, Navarro L. Electrochemical protoplast fusion in citrus. Plant Cell Rep 2005; 24(2): 112-9.
[http://dx.doi.org/10.1007/s00299-005-0916-1] [PMID: 15703946]

[116] Bengochea T, Doods JH. A plant protoplast: a biotechnological tool for plant improvement. London: Chapman & Hall 1986.
[http://dx.doi.org/10.1007/978-94-009-4095-6]

[117] Potrikus I, Shillito RD. Protoplasts: isolation, culture, plant regeneration. In: Weissbach A, Weissbach H, Eds. Methods for plant molecular biology. New York: Academic Press 1988; pp. 355-83.
[http://dx.doi.org/10.1016/B978-0-12-743655-5.50029-4]

[118] Evans DA, Cocking EC. Isolated plant protoplasts. In: Street HE, Ed. Plant tissue and cell culture. Oxford: Blackwell Scientific 1977; Vol. 11: pp. 103-35.

[119] Wang J, Jiang J, Wang Y. Protoplast fusion for crop improvement and breeding in China. Plant Cell Tissue Organ Cult 2012.
[http://dx.doi.org/10.1007/s11240-012-0221-y]

[120] Gamborg OL, Constabel F, Fowke L, *et al.* Protoplast and Cell culture methods in somatic hybridization in higher plants. Can J Genet Cytol 1974; 16: 737-50.
[http://dx.doi.org/10.1139/g74-080]

[121] Pental D, Cocking EC. Some theoretical and practical possibilities of plant genetic manipulation using protoplasts. Hereditas suppl 1985; 3: 89-92.
[http://dx.doi.org/10.1111/j.1601-5223.1985.tb00753.x]

[122] Wang YP, Sonntag K, Rudloff E. Development of rapeseed with high erucic acid content by asymmetric somatic hybridization between *Brassica napus* and *Crambe abyssinica*. Theor Appl Genet 2003; 106(7): 1147-55.
[http://dx.doi.org/10.1007/s00122-002-1176-x] [PMID: 12687349]

[123] Jiang JJ, Zhao XX, Tian W, Li TB, Wang YP. Intertribal somatic hybrids between *Brassica napus* and *Camelia sativa* with high linoleic acid content. Plant Cell Tissue Organ Cult 2009; 99: 91-5.
[http://dx.doi.org/10.1007/s11240-009-9579-x]

[124] Zhao ZG, Hu TT, Ge XH, Du XZ, Ding L, Li ZY. Production and characterization of intergeneric somatic hybrids between *Brassica napus* and *Orychophragmus violaceus* and their backcrossing progenies. Plant Cell Rep 2008; 27(10): 1611-21.
[http://dx.doi.org/10.1007/s00299-008-0582-1] [PMID: 18626647]

[125] Vasic DM, Taski KJ, Nagl NM, Skoric DM. Towards Sclerotinia resistance: somatic hybridization between wild and cultivated sunflower. Proceedings of the 16th International Sunflower Conference. Fargo, ND, USA. 2004; pp. 737-40.

[126] Vasil V, Wang DI, Vasil IK. Plant regeneration from protoplasts of napier grass (*Pennisetum purpureum* Schum.). Zeitchrift für Phlanzenphysiologie 1983; 111: 233-9.
[http://dx.doi.org/10.1016/S0044-328X(83)80082-8]

[127] Tudses N, Premjet S, Premjet D. Establishment of method for protoplast fusion with PEG-mediated between *Jatropha curcas* L. and *Ricinus communis* L. Int J Life Sci Biotech Pharma Res 2014; 4(1): 50-6.

[128] Tudses N, Premjet S, Premjet D. Optimal conditions for high-yield protoplast isolations of *Jatropha curcas* L. and *Ricinus communis* L. American-Eurasian J Agricultr Environ Sci 2014; 14(3): 221-30.

[129] Aftab F, Zafar Y, Iqbal J. Optimization of condition for electrofusion in sugarcane protoplast. Pak J Bot 2002; 34(3): 297-301.

[130] Liu F, Ryschka U, Marthe F, Klocke E, Schuman Z, Zao H. Culture and fusion of pollen protoplasts of *Brassica oleracea* L. var. italic with haploid mesophyll protoplasts of *B. rapa* L. spp. Pekinensis. Protoplasma 2007.
[http://dx.doi.org/10.1007/s00709-006-0228-5]

CHAPTER 3

DNA Repair: Its Molecular Basis for Use in Biotechnology

Felipe Augusto Godoy* and **Mateus Prates Mori**

University of São Paulo, Dept. of Biochemistry, Institute of Chemistry, São Paulo, Brazil

Abstract: All living organisms are constantly exposed to DNA-damaging agents, leading to the accumulation of chemical and structural modifications which can affect key processes such as replication and transcription. Various DNA repair pathways have evolved for dealing with these modifications, thus preventing their toxic and mutational potential. Mutations in mtDNA are frequently observed in several diseases, which are reflected in metabolic changes or even in the attenuation of apoptotic response to anticancer therapies. To the integrity of the mitochondrial genome, repair mechanisms are recruited to the organelle. Among these, the BER pathway is the main pathway localized to mitochondria. The identification of mechanisms that prevent the accumulation of DNA damage in plants with agricultural interest can lead to improvements of these crops, including sugarcane, thus leading to improvements in sugarcane processing and consequently in the production of biofuels.

Keywords: Biochemistry, BER pathway, Bioenergy, Biofuel, Biotechnology, DNA, DNA damage, DNA repair, DRR pathway, LP-BER, Mitochondria, MtBER, Mutations, NER pathway, Oligonucleotides, Oxidative damage, Plants, Repair enzymes, SP-BER, Sugarcane.

INTRODUCTION

"We totally missed the possible role of ... [DNA] repair although ... I later came to realize that DNA is so precious that probably many distinct repair mechanisms would exist."

Francis Crick, writing in Nature, 26 April 1974 [1].

With this quote, we would like to introduce the reader to a very important aspect of genome stability in every single living organism: DNA repair mechanisms. The acknowledgment by the scientific community of the relevance of these "ancient

---

* **Corresponding author Felipe Augusto Godoy:** Institute of Chemistry, University of Sao Paulo, Sao Paulo-SP, Brazil; Tel: +55 19 982207540; E-mail: felipebioquimica@yahoo.com.br

Daniela Defavari do Nascimento, & William A. Pickering (Eds.)

molecular copyists" yielded the 2015 Nobel Prize in chemistry to Thomas Lindahl, Aziz Sancar, and Paul Modrich. The copyist analogy seems suitable in terms of the replication and maintenance of original genetic information, even when taking into consideration that stochastic genetic alteration is regarded as one of the core dogmas of biological evolution.

From the natural point of view, there is a dialectic conflict between replication error tolerance with consequent mutation-driven evolution, and replication nonsense error with consequent annihilation of information existence. DNA repair regulates the rates of random mutation, postponing widespread information decay, favoring control, and skewing the effects of randomness.

In 1949, Renato Dulbecco [2] and Albert Kelner [3] independently reported the first DNA repair mechanism in bacteriophage and in the gram-positive bacteria *Streptomyces*, a phenomenon now known as photoreactivation. However, neither researcher characterized the mechanism, nor was it their intention to do so.

Molecular mechanisms could be finally addressed after Beukers and Berends [4], Richard Setlow [5, 6], and Wulff and Fraenkel [7] described UV-induced thymine dimer formation in DNA. There was subsequently a flourishing in research on DNA damage and repair. Rupert led the research on the enzymatic properties of photoreactivation [8 - 11]. Hanawalt, Carrier, Howard-Flanders, and Boyce led the research on the "dark repair mechanism", which was later described as excision repair [12 - 16].

DNA repair as we know it today can be divided didactically into eight distinct molecular pathways: a) direct reversal repair (DDR); b) base excision repair (BER); c) nucleotide excision repair (NER); d) mismatch repair (MMR); e) homologous recombination repair (HRR); f) non-homologous end joining (NHEJ); g) translesion synthesis (TLS); and h) DNA damage response (DDR) [17]. These pathways deal with the vast chemically distinct DNA modifications. Considering only the nitrogenous bases of DNA, they can give rise to the following DNA modifications: 1) oxidation, alkylation, or deamination driven; 2) UV-photoproducts; 3) covalent adducts with nitroamines, polycyclic aromatic hydrocarbon epoxides, reactive aldehydes, ketoaldehydes, and lipid peroxides; and 4) DNA protein, inter- and intrastrand crosslink. Furthermore, DNA modification in $2'$-deoxyribose ($2'$-dR) moiety brings forth: i) abasic sites; ii) single-strand breaks (SSB); and iii) double-strand breaks (DSB). Finally, and also important, DNA replication mechanisms themselves are not intrinsically flawless, as they can generate: a) base pair mismatches (non-canonical); b) DNA loops due to small insertions or deletions; and c) DSB due to premature replication fork collapse [18, 19].

The core logic of DNA repair pathways, with the exception of DRR, NHEJ, and TLS, consists of four steps: i) lesion recognition (and verification, depending on the pathway); ii) lesion excision; iii) repair synthesis; and iv) DNA ligation. DDR, NHEJ, and TLS may execute some of these steps, but not all of them. In the next section, the DRR, BER, and NER pathways will be briefly described. The DNA repair pathways were chosen according to the most relevant types of DNA damage that are inflicted in Plantae organisms, *i.e.,* UV-induced, alkylation, and oxidative DNA damage. We will discuss DNA repair pathways with regard to the eukaryote model, as well as introducing the historical importance of the prokaryote model *Escherichia coli* in gene discovery.

DIRECT REVERSAL REPAIR (DRR)

Direct reversal repair is the simplest DNA repair pathway. This is due to the unique feature of recognition and reversal of DNA damage by the same protein. There are three types of DDR proteins: i) DNA photolyases; ii) DNA alkyltransferases; and iii) oxidative demethylases of DNA.

DNA photolyase was the first enzyme attributed to a DNA repair phenomenon, and it was reported in the late 1940s [2, 3]. This molecular event was called photoreactivation at the time. Very briefly, photoreactivation is the reversal of the effects of UV on DNA by exposure of an organism to near-UV [20]; it was later discovered to be catalyzed by DNA photolyases.

DNA photolyases are closely related to cryptochromes, a family of flavoproteins largely distributed across eukaryotes responsible for blue-light-mediated plant development and mammalian circadian rhythm [21]. DNA photolyases possess a reduced flavin adenine dinucleotide ($FADH_2$), and a folate or deazaflavin [22]. In the dark (*i.e.,* light-independently), DNA photolyases bind with high affinity and specificity to pyrimidine dimers on dsDNA, a DNA lesion that arises from far-UV irradiation (200-300 nm). They absorb near-UV (300-500 nm) photons in order to split pyrimidine dimers back into two regular pyrimidines.

DNA photolyases are ubiquitous in almost every living organism [23]. The exception to this rule are the placental mammals, which, of course, include humans. The accepted proposition is that placental mammals lost DNA photolyase genes after switching from diurnal to nocturnal activity, a change which exercised low selective pressure on this class of genes.

The second class of DDR proteins are the DNA alkyltransferases. Some alkylation damage is repaired by direct removal [24]. These alkylation-induced DNA modifications – more specifically methylation – are directly reverted by the transferring of the extra methyl group in the DNA base to a cysteine residue in the

DNA alkyltransferase. Because the methylated cysteine residue cannot be eliminated from the protein, the whole polypeptide is therefore degraded. This is the reason why DNA alkyltransferases are called suicidal enzymes or suizymes [25]. There are basically two alkyltransferases: i) fused DNA-binding trans-criptional dual regulator/O^6-methylguanine-DNA methyltransferase (*ada*); and ii) ogt O^6-alkylguanine-DNA:cysteine-protein methyltransferase (*ogt*).

Oxidative demethylases of DNA lesions, AlkB, also perform DNA repair of some methylated bases. This class of enzymes uses α-ketoglutarate, O_2, and Fe^{2+} in order to attack the extra methyl group in the DNA base. Initially, AlkB oxidizes the methyl group in 1-methyladenine or 3-methylcytosine, yielding a hydroxy-methyl form, with subsequent decarboxylation of α-ketoglutarate to succinate. The hydroxymethyl group is then eliminated as a formaldehyde molecule, regenerating the unmodified base [26].

Concerning DDR encoding genes, *E. coli* has four (*phr*, *ada*, *ogt*, and *alkB*), and placental mammals have three (*ALKBH2*, *ALKBH3*, and *MGMT*). The plant model *Arabidopsis thaliana* genome possesses four gene coding DNA photolyases with different or overlapping substrate specificity: i) photolyase 1 (PHR1); ii) photolyase/blue-light receptor 2 (PHR2); iii) (6-4)DNA photolyase (UVR3); and iv) DNA photolyase (AT4G25290). It is likely that DNA photolyases are enriched in plants and, more broadly, in photosynthetic organisms, due to constant exposure to sunlight. To date, no alkyltransferase nor AlkB were described in *Arabidopsis*, but there are several AlkB orthologs in corn and soybean genomes.

BASE EXCISION REPAIR (BER)

The BER pathway deals mainly with small nucleobase DNA modifications, such as oxidation, alkylation, and deamination. Thomas Lindahl first isolated an *E. coli* enzyme that could release uracil from U:G mismatches in double-stranded DNA [27]. Many other components of the BER pathway, such as APE1 and DNA ligase 1, were also described by Lindahl in the 1970s [28, 29]. In 1979, Hayakawa and Sekigushi [30], and Lindahl [31], independently used the initial step of base elimination as an eponym for BER, differentiating two excision repair pathways, BER and NER. Finally, in 1994, Dianov and Lindahl reconstituted the BER pathway *in vitro* [32].

DNA repair *via* the BER pathway is enzymatically simple, consisting of five reactions catalyzed by four core proteins. Initially, a DNA glycosylase recognizes a modified DNA nucleobase and cleaves the *N*-β-glycosidic bond between the base and the 2'-dR, creating an apurinic/apyrimidic (AP or abasic) site. The remaining AP site undergoes endonucleolytic cleavage by AP endonuclease 1

(APE1) upstream of the AP site (*i.e.,* at the 5′ end), giving rise to a 3′-hydroxyl end (3′-OH) and a 2′-deoxyribose-5′-phosphate (5′-dRP). DNA polymerase β (Pol β) removes the 2′-dRP overhang and fills the gap with the complementary nucleotide. Finally, DNA Ligase III (Lig3), in complex with the auxiliary protein XRCC1, seals the cut in an ATP-dependent reaction (Fig. **1**).

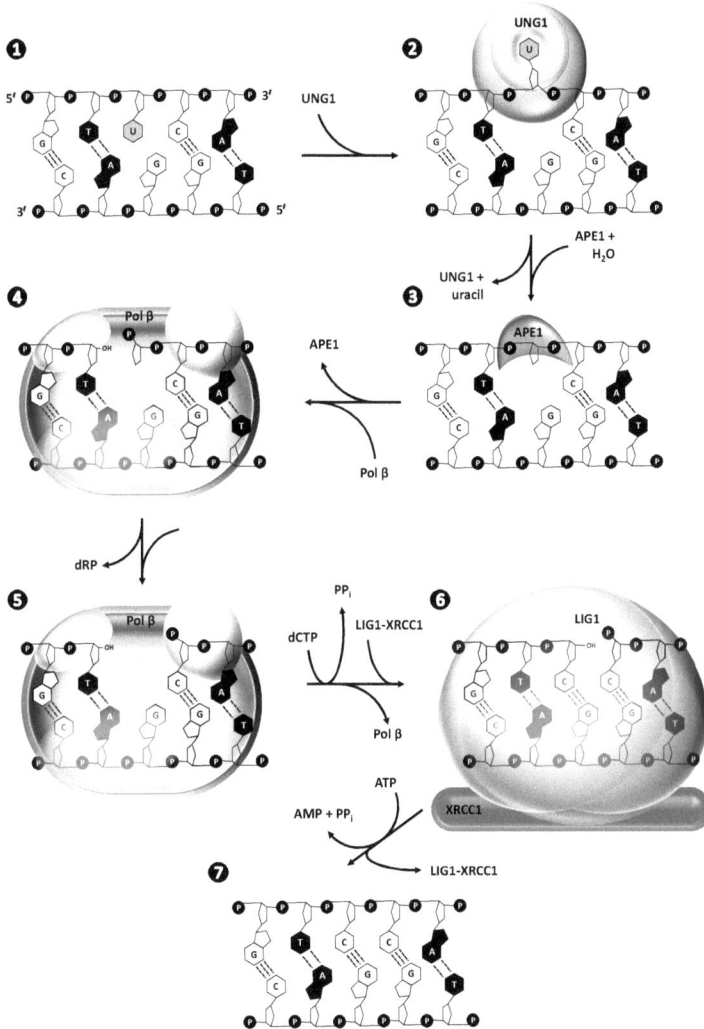

Fig. (1). Short patch base excision repair (SP-BER) scheme.
1) Cytosine deamination gives rise to a uracil:guanine (U:G) mismatch. 2) The U:G mismatch is recognized by UNG1, which flips-out U into its catalytic active site pocket and cleaves the N-β-glycosidic bond, releasing free U. 3) APE1 dislodges UNG1 from DNA and cuts the phosphodiester bond upstream of the AP site *via* a hydrolytic reaction mechanism, leaving a 3-OH end and a 5′-dRP. 4) POLβ dRP lyase activity trims the dRP end, generating a ligatable 5′-P. 5) POLβ inserts the complementary C opposite to G. 6) LIG1-XRCC1 seals the nick at the cost of one ATP. 7) DNA is repaired.

Aside from these basic steps, BER can be divided into two pathways according to the length of the newly synthesized strand: i) single or short-patch BER (SP-BER) and ii) long-patch BER (LP-BER). The steps in SP-BER have already been described above. The distinction of the LP-BER sub-pathway lies in the polymerization step. Pol β adds 2 to 11 nucleotides, displacing the downstream strand after APE1 processing. The displaced strand forms a flap-like structure, which is not a substrate for dRP lyase activity of Pol β. The flap endonuclease 1 (FEN1) cuts off the displaced strand upstream from the point of contact of the last base pairing. The cut is then ligated by DNA ligase I (Lig1), which does not require an auxiliary protein.

The recognition step commenced by the DNA glycosylase seems to be rate-limiting. DNA glycosylases vary in substrate-specificity and efficiency. Types of DNA glycosylases are based on the mechanism of catalysis: i) monofunctional DNA glycosylases, which have only DNA glycosylase activity and utilize a water molecule in order to attack the *N*-β-glycosdic bond; and ii) bifunctional DNA glycosylases, which display associated 3'-AP lyase activity.

Because the BER pathway is probably the most conserved DNA repair pathway, DNA glycosylases throughout the three domains show some orthology and, in a few cases, homology. The present data supports eight DNA glycosylases in *E. coli* and eleven in mammalians. Only four DNA glycosylases have been described in *Arabidopsis thaliana* so far: i) uracil DNA glycosylase (UNG); ii) 8-oxoguanine DNA glycosylase (OGG1); iii) the mutM homolog, formamidopyrimidine-DNA glycosylase (MMH-1); iv) endonuclease III 2 (NTH2); and v) DNA---methyladenine glycosylase (AT3G12040).

NUCLEOTIDE EXCISION REPAIR (NER)

NER is unanimously acknowledged as the DNA repair pathway possessing the greatest substrate plasticity [33]. In part, this feature is enabled by the molecular mechanism of lesion recognition in DNA, which is based on recognition of structural distortions of the DNA double helix induced by some base and strand modifications. The list of substrates of the NER pathway is chemically diverse and non-related, including: i) UV-induced DNA modifications (CPD and 6-4PP); ii) bulky adduct with activated polycyclic aromatic hydrocarbons; iii) oxidatively-induced lesions, such as cyclopurines; and iv) intrastrand adducts with chemotherapeutic drugs, such as cisplatin [34].

Previous to thorough NER description, Howard-Flanders and coworkers identified three excision repair loci in *E. coli*, namely *uvrA*, *uvrB*, and *uvrC* [35]. A significant causal link cast light on DNA repair when James E. Cleaver associated defective DNA repair in patients with xeroderma pigmentosum, a cancer-prone autosomal inherited disease [36]. However, it was only in 1981 that Aziz Sancar identified the three gene products through cloning [37 - 39]. *In vitro* reconstitution of eukaryotic NER requires twenty-five to thirty polypeptides, and was accomplished in 1995 by Aboussekhra and coworkers [40].

In short, this pathway can be understood in four steps: i) lesion recognition; ii) double helix unwinding and lesion verification; iii) pre-incision complex assembly; and iv) dual incision and repair synthesis. Furthermore, NER is divided into two sub-pathways, corresponding to the mechanisms of lesion recognition: i) a transcription-coupled repair sub-pathway (TC-NER) that repairs DNA lesions in template strands of actively transcribed genes, leading to increased readiness in repair; and ii) a global genome repair sub-pathway (GG-NER) that removes lesions present elsewhere in the genome. These sub-pathways diverge only at the lesion recognition step, converging in subsequent common events.

Whenever RNA polymerase I or II (RNA Pol) meets a DNA lesion on the template strand which blocks transcription progression, a possible molecular outcome is TC-NER. CSB protein (Cockayne syndrome complementation group B) is the first to respond to RNA Pol stalling, recruiting the CSA protein (Cockayne syndrome complementation group A). CSA is part of the E3 ubiquitin ligase complex with CUL4A (Cullin 4A), ROC1 (regulator of cullins 1), DDB1 (damage-specific DNA binding protein 1), and CSN (COP9 signalosome) [41]. This complex ubiquitinates CSB, which is necessary to fully resolve DNA damage allowing transcription resumption.

The GG-NER sub-pathway starts with lesion recognition by two sensor protein complexes: i) UV-DDB, a heterodimer composed of DDB1 and DDB2; and ii) the heterotrimer XPC-RAD23B-centrin2. Although UV-DDB facilitates lesion recognition, this complex is dispensable in GG-NER, acting more as an auxiliary protein to CPD repair. The XPC complex is the major DNA lesion recognition protein, due to its ability to sense small thermodynamic fluctuations of hydrogen bonding when base pairing is compromised due to the presence of the lesion [42].

After lesion recognition, TFIIH (transcription factor II Human) is recruited to the locale of the lesion. This multiprotein complex, which is also a canonical transcription factor and an essential component of NER pathways, is composed of ten subunits: i) a core composed of XPB, p52, p8, p62, p34 and p44; ii) a CAK complex (cyclin activated kinase) composed of CDK7 (cyclin-dependent kinase 7), cyclin H, and MAT1; and iii) XPD protein bridging core and CAK complex [41]. Both XPB and XPD are ATP-dependent DNA helicases, being responsible for unwinding the double helix in opposite polarities. XPD is the most probable candidate for the lesion verification step in NER [43 - 45].

Next, the pre-incision complex assembly occurs, with the binding of XPA, RPA, and XPG proteins, while the XPC complex leaves the repairsome. XPA is a core component of the pathways, interacting with several NER proteins such as subunits of TFIIH complex, RPA, XPC-RAD23B, DDB2, ERCC-XPF, and PCNA. This protein binds to the 5′ end of the damaged strand, near XPD (opposite to XPB). The RPA heterotrimer binds to the undamaged strand covering approximately thirty nucleotides, dislodging XPC that leaves the repair complex. Moreover, RPA assists the correct positioning of the two endonucleases, XPG and ERCC1-XPF, which are liable to cut the damage-containing oligonucleotide [46].

The damage-containing oligonucleotide incision does not happen until ERCC1-XPF arrival. Indeed, XPG associates with the pre-incision complex before ERCC1-XPF and its arrival is necessary for ERCC1-XPF recruitment. However, the latter is the first endonuclease to perform one incision in the damaged strand. The 3′ incision relative to the damaged site gives rise to a 3′-OH terminus that is used as a primer to DNA repair synthesis. Following the first incision, XPG cleaves the 5′ end at a 22-30 nucleotide distance from the ERCC1-XPF incision [47]. Oligonucleotide is then released, associated with TFIIH, while proteins from the incision complex dissociate from the repair site. Finally, canonical replication factors such as Pol δ, ε, or κ, PCNA, RFC, RPA, and DNA ligase are recruited to perform DNA repair synthesis [48, 49]. Once the DNA polymerization has ended, Lig1 or XRCC1-Lig3 complex seals the residual nick, fixing the DNA.

Eukaryotic NER is much more complex and requires many more proteins than prokaryotic NER. *E. coli* NER proteins are multitasking, while mammalian proteins are more single task performers. *A. thaliana* possesses two helicases homolog to human XPB (AtXPB1 and AtXPB2), which partially complement yeast rad25 mutant sensitivity to UV [50, 51]. Furthermore, *A. thaliana* has two DDB1 orthologs (DDB1A and DDB1B), the latter being essential to embryo development as a part of several E3 ubiquitin ligases [52]. In addition, both of them are involved in photomorphogenesis due to their interaction with COP1 (constitutively photomorphogenic 1), a conserved Ring finger E3 ubiquitin ligase that targets several photomorphogenesis-promoting proteins for destruction [53]. DDB1B, DDB2, and UVH6 defective *A. thaliana* are UV hypersensitive. DDB1A acts in the recognition step of UV-induced DNA lesions, as well as DDB2 [54]. In turn, UVH6 is a homolog of human XPD helicase and exerts the same exact function in *A. thaliana* NER. UVH1 (or AtRAD1) is the *A. thaliana* homolog of human XPF, and forms a complex with AtERCC1 [55]. Indeed, UVH1 deficient *Arabidopsis* are also UV hypersensitive [56].

BASIC TECHNIQUES AND APPLIED FUNDAMENTALS

All living organisms are constantly exposed to a variety of agents that can cause chemical and/or structural changes in DNA, affecting essential processes such as replication and transcription. Throughout evolution, several DNA repair strategies have developed to remove these changes, preventing toxic or mutagenic effects of these injuries [57]. To maintain mitochondrial genomic integrity, some repair mechanisms are recruited to the organelle, especially repair *via* base excision, BER [58]. Mitochondrial BER (mtBER), which uses a similar set of nuclear BER proteins and biochemical process, is similar except for the presence of DNA γ polymerase exclusively in mitochondria that are involved in both replication and repair [59].

In this context, the following sections of this chapter describe procedures for measuring the formation of oxidative damage and mtBER, including mitochondria isolation from tissues and cells and determination of mtBER enzyme activity. These sections therefore propose to provide tools for the investigation of repair mechanisms.

ISOLATION OF MITOCHONDRIA

For DNA repair assays, it is important to minimize nuclear contamination and maintain the activity of repair enzymes, because the existence of a very large nuclear contamination in mitochondrial fraction consists of an artifact that masks functions and activities in mitochondrial fractions. It is essential to freeze the tissue immediately after sacrificing animals and harvesting plant tissues. The enzymatic activity is maintained by performing all steps in ice, centrifugation at 4°C, and by the addition of protease inhibitors, all in an attempt to minimize digestion [60].

MITOCHONDRIAL ISOLATION OF TISSUES AND CULTURED CELLS

Most tissues may be isolated following the methods already described [60, 61]. The tissue placed in MSHE buffer at 4°C, where the tissue is triturated in this buffer with the aid of scissors and washed with MSHE. This tissue is homogenized until it has a smooth look. This homogenate is centrifuged at a rate of 1000g for ten minutes, where nuclear fractions and whole cells are precipitated. By centrifuging supernatant at 10000g for ten minutes, the obtained pellet contains mitochondrial fraction which must then be washed with MSHE, followed by another centrifugation at 10000g for ten minutes.

The final pellet is resuspended in 2x MSHE buffer and 50% *Percoll* in a ratio of 1:1, and subjected to centrifugation 50000g for 120 minutes. The mitochondrial fraction appears as the third band from the top of the centrifuge tube. Mitochondria are removed and washed by addition of ten volumes of buffer MSHE, an important procedure for getting rid of *Percoll*. Pure mitochondrial extract is precipitated in a centrifuge at 3000g for ten minutes.

In general, mitochondria can be isolated, requiring small protocol adjustments, depending on the tissues in question. For example, in brain tissue a portion of mitochondria is contained in synaptosomal vesicles, and the addition of digitonin, a detergent, could assist the release of these vesicles, increasing the yield. In the case of plant tissues, enzymes must be used to break cell walls.

When it comes to cell culture, the limiting factors in these cases are cell type and quantity. There must be between 30 and 150 mm tissue on a culture plate, either animal or vegetable, to be sufficient for obtaining a significant amount of mitochondria [60].

At this point, it is important to check the purity of the mitochondrial preparation. One widely used technique is *western blotting* with the aid of a nuclear contamination probe, lamin B2, which is a fairly abundant nuclear protein.

MITOCHONDRIAL EXTRACTS

From mitochondrial fraction, an extract should be prepared for *in vitro* assays. Lysis buffer is added by ninety minutes incubation at 4°C. The lysate is sonicated, five times for five seconds, on ice with intervals of thirty seconds, and centrifuged at 130000g to precipitate the DNA and membrane contaminants. The supernatant contains the purified protein [60].

The protein concentration is measured and assays may be performed. It is noteworthy that these aliquots may be stored at -80°C for several months without significant loss of enzyme activity.

MTBER ENZYME ACTIVITY

The molecular mechanism of the BER occurs in four biochemical steps [58]:

1) It starts with the recognition and excision of the modified base by the DNA glycolass, which generates an abasic site (AP).

2) The AP site is then recognized by the other group of enzymes, AP-endonuclease, which make an incision at the 3' or 5' of the AP site, generating single-strand breaks.

3) After removal of the terminal deoxyribose-5' phosphate by repair DNA

polymerase (usually DNA polymerase β in the nucleus and γ in the mitochondria), there is the insertion of a nucleotide by the same enzyme.

4) The strand is finally linked by DNA ligase.

Each of these biochemical steps can be measured separately or as a whole by means of an incorporated assay. There are several ways to perform these measurements, but the most commonly used method is based on oligonucleotides. One of the major advantages of oligonucleotide-based assays is that it is possible to design specific substrates, allowing specific activity probes of each enzyme in individual steps [60].

DNA GLYCOSYLASE ASSAY

DNA glycosylase catalyzes the first reaction in the BER pathway, excising the damaged base by cleavage of the N-glycosyl bond, linking the base to the ribose of the nucleotide. These enzymes recognize specific base changes, and as such have a defined substrate specificity [60].

Tests to measure the activity depend on their activity in decomposing the oligonucleotide containing the lesion. The substrates are then resolved in a denaturing polyacrylamide gel, and activity is calculated as the amount of substrate cleaved in the total amount of substrate in the control group [60].

AP ENDONUCLEASE ASSAY

The next step of the BER pathway is the cleavage of the resulting abasic site of the DNA glycosylase activity. In mammals, both in the nucleus and in the mitochondria, endonuclease AP1 (APE1) is the main endonuclease that catalyzes this step. The test is similar to the incision glycosylases, but the substrate already contains a true abasic site generated through a reaction of class I DNA glycosylases [60].

NUCLEOTIDE INSERT BY POLIMERASE γ ASSAY

The next step in the BER pathway is the removal of the terminal deoxyribose-5' phosphate by the repair DNA polymerase (usually DNA polymerase β in the nucleus and γ in the mitochondria) and insertion of a nucleotide by the same enzyme [60]. The insert activity is measured using an unlabeled oligonucleotide containing one abasic site, and the assay measures the incorporation of nucleotides during repair [60].

REPAIR AND INCORPORATION ASSAYS

The repair and incorporation assays measure the BER pathway activity as a whole, in which the incorporation of a radioactive nucleotide in a new substrate containing damage is quantitated [60].

BER activity is quantified as the intensity of the 32P-dCTP signal bandwidth product compared to a control sample, after subtracting the background of a reaction without protein [60].

All the assays described here require the resolution of the oligonucleotides under denaturing conditions, and the signals are then obtained using a phosphorimager scanner or equivalent. The radioactivity on the columns is quantified using the ImageQuant software [60].

SINGLE-CELL GEL (SCG)

When at an advanced stage, after the repair system does not work, genomic DNA is fragmented due to the activation of enzymes, and the waste can be detected through the single-cell gel test, or "comet assay" [62].

After incubation/treatment, cells are centrifuged and resuspended in phosphate buffer. To ensure the viability of the cells, a trypan blue exclusion test is performed. Only samples with viability above 80% should be analyzed. Slides are covered with the mixture of cells with LMP agarose and covered with coverslips. After solidification, the coverslips are removed, and the blades are placed in lysing solution with DMSO and Triton X-100, and refrigerated for 24 hours. The blades are then placed for twenty minutes in electrophoresis solution to trigger the cleavage of DNA unfolding and alkali-labile sites. Thereafter, electrophoresis is performed for twenty minutes at 25 V and 300 mA. Then the slides are placed for twenty minutes in a neutralizing solution and subsequently dehydrated in ethanol at room temperature for five minutes and stained with DAPI. Cells are examined by random sampling and visualized by a fluorescence microscope with a 358/461 nm filter for absorption/emission.

MEASUREMENT OF MTDNA INJURIES THROUGH LONG-EXTENSION PCR

Long-extension PCR (LX-PCR), followed by southern hybridization with probes for two different regions of the mitochondrial genome, is widely used to evaluate the presence of deleted mtDNA molecules [63].

For amplification of mtDNA fragment (13.4 kb), the thermal cycler is programmed to initial denaturation at 94°C for five minutes, eighteen cycles of

94°C for 30 seconds, and 68°C for 12.5 minutes, with a final extension at 72°C for ten minutes. The DNA damage will be quantified by comparing the relative efficiency of amplification of large DNA fragments (13.4 kb mitochondrial DNA), and normalizing them to the amplification of smaller fragments (235 bp and 195 bp) [64].

HOST CELL REACTIVATION

Host cell reactivation (HCR), was first used to describe the survival of UV-irradiated bacteriophages that were transfected to UV-pretreated cells. HCR measure the DNA repair capacity of a DNA alteration in the cell [65].

The host cell is transfected with a damaged plasmid that contains a reporter gene, deactivated due to the damage. The ability of the cell to repair the damaged plasmid, after it is introduced to the cell, will allow the reactivation of the reporter gene [66].

CONCLUSION AND FUTURE PROSPECTS

When it comes to plants, there are many varieties of vegetables with economic importance. However, sugarcane has had a very high profile due to biofuel production from its raw material. Sugarcane is one of the most important energy crops, and there is great interest in developing more resistant plants [67, 68].

In this context, a Brazilian research group has sequenced a large number of sugarcane cDNA expressed sequence tags (ESTs). This project, the Sugarcane Expressed Sequence Tag (SUCEST) genome project, provides an extensive database of different EST libraries [67 - 69].

Identification of genes involved in plant DNA repair is of great interest, because of its importance in maintaining genome integrity, thus preventing the disruption of essential cellular processes such as DNA replication and transcription. Plants are constantly exposed to damaging environmental agents because of their sessile life style, but very little is known about DNA repair systems in plants [67, 69, 70].

Avoiding the accumulation of DNA damage in plants can lead to agricultural improvements in various crops, such as sugarcane [67 - 71]. It may also lead to an understanding of the genes absent from plants. Some of these indicate important differences in DNA repair mechanisms, leading to an improvement in sugarcane processing and consequently the production of biofuels.

CONFLICT OF INTEREST

The authors confirm that this chapter content has no conflict of interest.

ACKNOWLEDGEMENTS

None declared.

REFERENCES

[1]   Crick F. The double helix: a personal view. Nature 1974; 248(5451): 766-9.
        [http://dx.doi.org/10.1038/248766a0] [PMID: 4599081]

[2]   Kelner A. Effect of visible light on the recovery of *Streptomyces griseus* conidia from ultra-violet
        irradiation injury. Proc Natl Acad Sci USA 1949; 35(2): 73-9.
        [http://dx.doi.org/10.1073/pnas.35.2.73] [PMID: 16588862]

[3]   Dulbecco R. Reactivation of ultra-violet-inactivated bacteriophage by visible light. Nature 1949;
        163(4155): 949.
        [http://dx.doi.org/10.1038/163949b0] [PMID: 18229246]

[4]   Beukers R, Berends W. Isolation and identification of the irradiation product of thymine. Biochim
        Biophys Acta 1960; 41: 550-1.
        [http://dx.doi.org/10.1016/0006-3002(60)90063-9] [PMID: 13800233]

[5]   Setlow R. The action spectrum for the reversal of the dimerization of thymine induced by ultraviolet
        light. Biochim Biophys Acta 1961; 49: 237-8.
        [http://dx.doi.org/10.1016/0006-3002(61)90894-0] [PMID: 13750384]

[6]   Setlow RB, Setlow JK. The proper use of short-wavelength reversal as a criterion of the importance of
        pyrimidine dimers in biological inactivation. Photochem Photobiol 1965; 4(5): 939-40.
        [http://dx.doi.org/10.1111/j.1751-1097.1965.tb07943.x] [PMID: 5297211]

[7]   Wulff DL, Fraenkel G. On the nature of thymine photo-product. Biochim Biophys Acta 1961; 51:
        332-9.
        [http://dx.doi.org/10.1016/0006-3002(61)90174-3] [PMID: 14008552]

[8]   Rupert CS. Repair of ultraviolet damage in cellular DNA J Cell Comp Physiol 1961; 58(3): 57-68.

[9]   Rupert CS. Photoenzymatic repair of ultraviolet damage in DNA. I. Kinetics of the reaction. J Gen
        Physiol 1962; 45: 703-24.
        [http://dx.doi.org/10.1085/jgp.45.4.703] [PMID: 14495308]

[10]   Rupert CS. Photoenzymatic repair of ultraviolet damage in DNA. II. Formation of an enzyme-
        substrate complex. J Gen Physiol 1962; 45: 725-41.
        [http://dx.doi.org/10.1085/jgp.45.4.725] [PMID: 14495309]

[11]   Wulff DL, Rupert CS. Disappearance of thymine photodimer in ultraviolet irradiated DNA upon
        treatment with a photoreactivating enzyme from bakers yeast. Biochem Biophys Res Commun 1962;
        7: 237-40.
        [http://dx.doi.org/10.1016/0006-291X(62)90181-X] [PMID: 14008553]

[12]   Howard-Flanders P, Boyce RP, Simson E, Theriot L. A genetic locus in *E. coli* K12 that controls the
        reactivation of UV-photoproducts associated with thymine in DNA. Proc Natl Acad Sci USA 1962;
        48: 2109-15.
        [http://dx.doi.org/10.1073/pnas.48.12.2109] [PMID: 13955137]

[13]   Boyce RP, Howard-Flanders P. Release of ultraviolet light-induced thymine dimers from DNA in *E.
        coli* K-12. Proc Natl Acad Sci USA 1964; 51: 293-300.
        [http://dx.doi.org/10.1073/pnas.51.2.293] [PMID: 14124327]

[14]   Shuster RC, Boyce RP. The excision of thymine dimer from the DNA of UV-irradiated *E. coli* 15 T-
        A-U during thymine deprivation. Biochim Biophys Res Commun 1964; 16(5): 489-96.
        [http://dx.doi.org/10.1016/0006-291X(64)90381-X] [PMID: 5332853]

[15]   Schuster RC. Dark repair of ultraviolet injury in *E. coli* during deprivation of thymine. Nature 1964; 202: 614-5.
       [http://dx.doi.org/10.1038/202614a0] [PMID: 14195082]

[16]   Pettijohn D, Hanawalt P. Evidence for repair-replication of ultraviolet damaged DNA in bacteria. J Mol Biol 1964; 9: 395-410.
       [http://dx.doi.org/10.1016/S0022-2836(64)80216-3] [PMID: 14202275]

[17]   Milanowska K, Krwawicz J, Papaj G, *et al.* REPAIRtoirea database of DNA repair pathways. Nucleic Acids Res 2011; 39(Database issue): D788-92.
       [http://dx.doi.org/10.1093/nar/gkq1087] [PMID: 21051355]

[18]   Sancar A, Lindsey-Boltz LA, Unsal-Kaçmaz K, Linn S. Molecular mechanisms of mammalian DNA repair and the DNA damage checkpoints. Annu Rev Biochem 2004; 73: 39-85.
       [http://dx.doi.org/10.1146/annurev.biochem.73.011303.073723] [PMID: 15189136]

[19]   Branzei D, Foiani M. Regulation of DNA repair throughout the cell cycle. Nat Rev Mol Cell Biol 2008; 9(4): 297-308.
       [http://dx.doi.org/10.1038/nrm2351] [PMID: 18285803]

[20]   Myles GM, Sancar A. DNA repair. Chem Res Toxicol 1989; 2(4): 197-226.
       [http://dx.doi.org/10.1021/tx00010a001] [PMID: 2519777]

[21]   Cashmore AR, Jarillo JA, Wu Y-J, Liu D. Cryptochromes: blue light receptors for plants and animals. Science 1999; 284(5415): 760-5.
       [http://dx.doi.org/10.1126/science.284.5415.760] [PMID: 10221900]

[22]   Sancar GB, Sancar A. Structure and function of DNA photolyases. Trends Biochem Sci 1987; 12: 259-61.
       [http://dx.doi.org/10.1016/0968-0004(87)90130-7]

[23]   Kanai S, Kikuno R, Toh H, Ryo H, Todo T. Molecular evolution of the photolyase-blue-light photoreceptor family. J Mol Evol 1997; 45(5): 535-48.
       [http://dx.doi.org/10.1007/PL00006258] [PMID: 9342401]

[24]   Samson L, Thomale J, Rajewsky MF. Alternative pathways for the *in vivo* repair of O^6-alkylguanine and O^4-alkylthymine in *Escherichia coli*: the adaptive response and nucleotide excision repair. EMBO J 1988; 7(7): 2261-7.
       [PMID: 3046938]

[25]   Lindahl T, Sedgwick B, Sekiguchi M, Nakabeppu Y. Regulation and expression of the adaptive response to alkylating agents. Annu Rev Biochem 1988; 57: 133-57.
       [http://dx.doi.org/10.1146/annurev.bi.57.070188.001025] [PMID: 3052269]

[26]   Trewick SC, Henshaw TF, Hausinger RP, Lindahl T, Sedgwick B. Oxidative demethylation by *Escherichia coli* AlkB directly reverts DNA base damage. Nature 2002; 419(6903): 174-8.
       [http://dx.doi.org/10.1038/nature00908] [PMID: 12226667]

[27]   Lindahl T. An N-glycosidase from *Escherichia coli* that releases free uracil from DNA containing deaminated cytosine residues. Proc Natl Acad Sci USA 1974; 71(9): 3649-53.
       [http://dx.doi.org/10.1073/pnas.71.9.3649] [PMID: 4610583]

[28]   Ljungquist S, Lindahl T. A mammalian endonuclease specific for apurinic sites in double-stranded deoxyribonucleic acid. I. Purification and general properties. J Biol Chem 1974; 249(5): 1530-5.
       [PMID: 4206354]

[29]   Söderhäll S, Lindahl T. Mammalian deoxyribonucleic acid ligase. Isolation of an active enzyme-adenylate complex. J Biol Chem 1973; 248(2): 672-5.
       [PMID: 4346342]

[30]   Hayakawa H, Sekiguchi M. [A new repair pathway which involves direct excision of bases from DNA (authors transl)]. Tanpakushitsu Kakusan Koso 1979; 24(5): 639-51. [A new repair pathway which involves direct excision of bases from DNA (author's translation)]. [PMID: 112647]

[31]   Lindahl T. DNA glycosylases, endonucleases for apurinic/apyrimidinic sites, and base excision-repair. Prog Nucleic Acid Res Mol Biol 1979; 22: 135-92. [http://dx.doi.org/10.1016/S0079-6603(08)60800-4] [PMID: 392601]

[32]   Dianov G, Lindahl T. Reconstitution of the DNA base excision-repair pathway. Curr Biol 1994; 4(12): 1069-76. [http://dx.doi.org/10.1016/S0960-9822(00)00245-1] [PMID: 7535646]

[33]   Schärer OD. Nucleotide excision repair in eukaryotes. Cold Spring Harb Perspect Biol 2013; 5(10): a012609. [http://dx.doi.org/10.1101/cshperspect.a012609] [PMID: 24086042]

[34]   Gillet LC, Schärer OD. Molecular mechanisms of mammalian global genome nucleotide excision repair. Chem Rev 2006; 106(2): 253-76. [http://dx.doi.org/10.1021/cr040483f] [PMID: 16464005]

[35]   Howard-Flanders P, Boyce RP, Theriot L. Three loci in *Escherichia coli* K-12 that control the excision of pyrimidine dimers and certain other mutagen products from DNA. Genetics 1966; 53(6): 1119-36. [PMID: 5335128]

[36]   Cleaver JE. Defective repair replication of DNA in xeroderma pigmentosum. Nature 1968; 218(5142): 652-6. [http://dx.doi.org/10.1038/218652a0] [PMID: 5655953]

[37]   Sancar A, Wharton RP, Seltzer S, Kacinski BM, Clarke ND, Rupp WD. Identification of the uvrA gene product. J Mol Biol 1981; 148(1): 45-62. [http://dx.doi.org/10.1016/0022-2836(81)90234-5] [PMID: 6273577]

[38]   Sancar A, Clarke ND, Griswold J, Kennedy WJ, Rupp WD. Identification of the uvrB gene product. J Mol Biol 1981; 148(1): 63-76. [http://dx.doi.org/10.1016/0022-2836(81)90235-7] [PMID: 6273578]

[39]   Sancar A, Kacinski BM, Mott DL, Rupp WD. Identification of the uvrC gene product. Proc Natl Acad Sci USA 1981; 78(9): 5450-4. [http://dx.doi.org/10.1073/pnas.78.9.5450] [PMID: 7029536]

[40]   Aboussekhra A, Biggerstaff M, Shivji MK, *et al.* Mammalian DNA nucleotide excision repair reconstituted with purified protein components. Cell 1995; 80(6): 859-68. [http://dx.doi.org/10.1016/0092-8674(95)90289-9] [PMID: 7697716]

[41]   Groisman R, Polanowska J, Kuraoka I, *et al.* The ubiquitin ligase activity in the DDB2 and CSA complexes is differentially regulated by the COP9 signalosome in response to DNA damage. Cell 2003; 113(3): 357-67. [http://dx.doi.org/10.1016/S0092-8674(03)00316-7] [PMID: 12732143]

[42]   Min J-H, Pavletich NP. Recognition of DNA damage by the Rad4 nucleotide excision repair protein. Nature 2007; 449(7162): 570-5. [http://dx.doi.org/10.1038/nature06155] [PMID: 17882165]

[43]   Compe E, Egly J-M. TFIIH: when transcription met DNA repair. Nat Rev Mol Cell Biol 2012; 13(6): 343-54. [http://dx.doi.org/10.1038/nrm3350] [PMID: 22572993]

[44]   Fan L, Fuss JO, Cheng QJ, *et al.* XPD helicase structures and activities: insights into the cancer and aging phenotypes from XPD mutations. Cell 2008; 133(5): 789-800. [http://dx.doi.org/10.1016/j.cell.2008.04.030] [PMID: 18510924]

[45]   Sugasawa K, Akagi J, Nishi R, Iwai S, Hanaoka F. Two-step recognition of DNA damage for mammalian nucleotide excision repair: Directional binding of the XPC complex and DNA strand scanning. Mol Cell 2009; 36(4): 642-53.
       [http://dx.doi.org/10.1016/j.molcel.2009.09.035] [PMID: 19941824]

[46]   de Laat WL, Appeldoorn E, Sugasawa K, Weterings E, Jaspers NG, Hoeijmakers JH. DNA-binding polarity of human replication protein A positions nucleases in nucleotide excision repair. Genes Dev 1998; 12(16): 2598-609.
       [http://dx.doi.org/10.1101/gad.12.16.2598] [PMID: 9716411]

[47]   Marteijn JA, Lans H, Vermeulen W, Hoeijmakers JH. Understanding nucleotide excision repair and its roles in cancer and ageing. Nat Rev Mol Cell Biol 2014; 15(7): 465-81.
       [http://dx.doi.org/10.1038/nrm3822] [PMID: 24954209]

[48]   Shivji MK, Podust VN, Hübscher U, Wood RD. Nucleotide excision repair DNA synthesis by DNA polymerase epsilon in the presence of PCNA, RFC, and RPA. Biochemistry 1995; 34(15): 5011-7.
       [http://dx.doi.org/10.1021/bi00015a012] [PMID: 7711023]

[49]   Araújo SJ, Tirode F, Coin F, *et al.* Nucleotide excision repair of DNA with recombinant human proteins: definition of the minimal set of factors, active forms of TFIIH, and modulation by CAK. Genes Dev 2000; 14(3): 349-59.
       [PMID: 10673506]

[50]   Costa RM, Morgante PG, Berra CM, *et al.* The participation of AtXPB1, the XPB/RAD25 homologue gene from *Arabidopsis thaliana*, in DNA repair and plant development. Plant J 2001; 28(4): 385-95.
       [http://dx.doi.org/10.1046/j.1365-313X.2001.01162.x] [PMID: 11737776]

[51]   Morgante PG, Berra CM, Nakabashi M, Costa RM, Menck CF, Van Sluys MA. Functional XPB/RAD25 redundancy in *Arabidopsis* genome: characterization of AtXPB2 and expression analysis. Gene 2005; 344: 93-103.
       [http://dx.doi.org/10.1016/j.gene.2004.10.006] [PMID: 15656976]

[52]   Bernhardt A, Mooney S, Hellmann H. *Arabidopsis* DDB1a and DDB1b are critical for embryo development. Planta 2010; 232(3): 555-66.
       [http://dx.doi.org/10.1007/s00425-010-1195-9] [PMID: 20499085]

[53]   Chen H, Huang X, Gusmaroli G, *et al. Arabidopsis* CULLIN4-damaged DNA binding protein 1 interacts with CONSTITUTIVELY PHOTOMORPHOGENIC1-SUPPRESSOR OF PHYA complexes to regulate photomorphogenesis and flowering time. Plant Cell 2010; 22(1): 108-23.
       [http://dx.doi.org/10.1105/tpc.109.065490] [PMID: 20061554]

[54]   Ly V, Hatherell A, Kim E, Chan A, Belmonte MF, Schroeder DF. Interactions between *Arabidopsis* DNA repair genes UVH6, DDB1A, and DDB2 during abiotic stress tolerance and floral development. Plant Sci 2013; 213: 88-97.
       [http://dx.doi.org/10.1016/j.plantsci.2013.09.004] [PMID: 24157211]

[55]   Li A, Schuermann D, Gallego F, Kovalchuk I, Tinland B. Repair of damaged DNA by *Arabidopsis* cell extract. Plant Cell 2002; 14(1): 263-73.
       [http://dx.doi.org/10.1105/tpc.010258] [PMID: 11826311]

[56]   Harlow GR, Jenkins ME, Pittalwala TS, Mount DW. Isolation of uvh1, an *Arabidopsis* mutant hypersensitive to ultraviolet light and ionizing radiation. Plant Cell 1994; 6(2): 227-35.
       [http://dx.doi.org/10.1105/tpc.6.2.227] [PMID: 8148646]

[57]   Friedberg EC. DNA damage and repair. Nature 2003; 421(6921): 436-40.
       [http://dx.doi.org/10.1038/nature01408] [PMID: 12540918]

[58]   Bohr VA, Stevnsner T, de Souza-Pinto NC. Mitochondrial DNA repair of oxidative damage in mammalian cells. Gene 2002; 286(1): 127-34.
       [http://dx.doi.org/10.1016/S0378-1119(01)00813-7] [PMID: 11943468]

[59]   de Souza-Pinto NC, Eide L, Hogue BA, *et al.* Repair of 8-oxodeoxyguanosine lesions in mitochondrial dna depends on the oxoguanine dna glycosylase (OGG1) gene and 8-oxoguanine accumulates in the mitochondrial dna of OGG1-defective mice. Cancer Res 2001; 61(14): 5378-81.
[PMID: 11454679]

[60]   Maynard S, de Souza-Pinto NC, Scheibye-Knudsen M, Bohr VA. Mitochondrial base excision repair assays. Methods 2010; 51(4): 416-25.
[http://dx.doi.org/10.1016/j.ymeth.2010.02.020] [PMID: 20188838]

[61]   Croteau DL, Bohr VA. Repair of oxidative damage to nuclear and mitochondrial DNA in mammalian cells. J Biol Chem 1997; 272(41): 25409-12.
[http://dx.doi.org/10.1074/jbc.272.41.25409] [PMID: 9325246]

[62]   Tice RR, Agurell E, Anderson D, *et al.* Single cell gel/comet assay: guidelines for *in vitro* and *in vivo* genetic toxicology testing. Environ Mol Mutagen 2000; 35(3): 206-21.
[http://dx.doi.org/10.1002/(SICI)1098-2280(2000)35:3<206::AID-EM8>3.0.CO;2-J] [PMID: 10737956]

[63]   Kajander OA, Kunnas TA, Perola M, Lehtinen SK, Karhunen PJ, Jacobs HT. Long-extension PCR to detect deleted mitochondrial DNA molecules is compromized by technical artefacts. Biochem Biophys Res Commun 1999; 254(2): 507-14.

[64]   Kovalenko OA, Santos JH. Analysis of oxidative damage by gene-specific quantitative PCR. Molecular Toxicology Protocols 2009; 291: 321-35.
[http://dx.doi.org/10.1002/0471142905.hg1901s62]

[65]   Johnson JM, Latimer JJ. Analysis of DNA repair using transfection-based host cell reactivation. Methods Mol Biol 2005; 291: 321-35.
[PMID: 15502233]

[66]   McCready S. An Immunoassay for Measuring Repair of DNA. Molecular Toxicology Protocols 2014; 1105: 551-64.
[http://dx.doi.org/10.1007/978-1-62703-739-6_38]

[67]   Costa RM, Lima WC, Vogel CI, *et al.* DNA repair-related genes in sugarcane expressed sequence tags (ESTs). Genet Mol Biol 2001; 24: 131-40.
[http://dx.doi.org/10.1590/S1415-47572001000100018]

[68]   Kanaar R, Hoeijmakers JH, van Gent DC. Molecular mechanisms of DNA double strand break repair. Trends Cell Biol 1998; 8(12): 483-9.
[http://dx.doi.org/10.1016/S0962-8924(98)01383-X] [PMID: 9861670]

[69]   Memisoglu A, Samson L. Base excision repair in yeast and mammals. Mutat Res 2000; 451(1-2): 39-51.
[http://dx.doi.org/10.1016/S0027-5107(00)00039-7] [PMID: 10915864]

[70]   Ries G, Heller W, Puchta H, Sandermann H, Seidlitz HK, Hohn B. Elevated UV-B radiation reduces genome stability in plants. Nature 2000; 406(6791): 98-101.
[http://dx.doi.org/10.1038/35017595] [PMID: 10894550]

[71]   Vonarx EJ, Mitchell HL, Karthikeyan R, Chatterjee I, Kunz BA. DNA repair in higher plants. Mutat Res 1998; 400(1-2): 187-200.
[http://dx.doi.org/10.1016/S0027-5107(98)00043-8] [PMID: 9685637]

CHAPTER 4

New Breeding Techniques

Humberto J. Debat[*]

Institute of Plant Pathology – Center of Agricultural Research – IPAVE – CIAP – INTA, Argentina

Abstract: A growing world population is demanding food and energy at the highest pace in the history of human kind. Plant biotechnology oriented to sustainable and competitive agriculture should lead a trail of innovation in order to address these emerging needs. The last decade has been characterized by the explosive accumulation of vast amounts of discoveries in the field of molecular biology. These achievements have paved the way for the advent of novel developments in plant biotechnology. Several new tools are being applied for genetic crop improvement every day, based on breakthrough versatile platforms that are increasing effectiveness and speed in the generation of new varieties of crops. The latest achievements are not only adaptations or improvements of modern techniques such as intra- and cis-genesis and accelerated breeding based on transgenics and null-segregants, meganucleases, and zinc finger nucleases; they are also the dawn of new disruptive technologies that are reshaping the paradigm of genetic improvement, as is the case with TALEN and CRISPR/Cas. Site-directed genome editing is becoming precise, cost-effective, versatile, and fast. These new breeding platforms are leading the way to next generation biotechnology. This chapter discusses the most recent updates and developments of new breeding techniques, paradigmatic achievements, future perspectives, and challenges in the context of plant biotechnology.

Keywords: Accelerated breeding, CRISPR/Cas, Genome editing, Intra-cis-genesis, Meganucleases, Oligonucleotide directed mutagenesis, Reverse breeding, Site-specific nucleases, Targeted mutagenesis, Transcription activator-like effectors nucleases, Zinc finger nucleases.

INTRODUCTION

The last decade has been marked by the constant emergence of breakthrough discoveries in molecular biology. We are experiencing a paradigm shift in experimental plant science, illustrated by the dynamic transformation of knowledge into innovative technological platforms and tools. New breeding

---

[*] **Corresponding author Humberto J. Debat:** Institute of Plant Pathology – Center of Agricultural Research – IPAVE – CIAP – INTA, Argentina; Tel: +543514973636; E-mail: debat.humberto@inta.gob.ar

Daniela Defavari do Nascimento, & William A. Pickering (Eds.)

techniques are emerging as incremental advances in traditional practices and as disruptive technologies that generate extraordinary progress at the fastest pace ever. This chapter will focus on the most innovative recent developments in a context of profound developmental shift. Several technologies devoted to crop improvement are presented, such as zinc finger nucleases (ZFN), transcription activator–like effectors nucleases (TALEN), and CRISPR/Cas 9, among others. The benefits and challenges of several technological processes and tools are discussed, highlighting some aspects related to the associated regulatory framework and focusing on the most advanced and promising technological developments.

Zinc Finger Nucleases

Shukla *et al*. [1] inaugurated the use of ZFNs (Fig. **1b**) for editing endogenous genes in plants. The process was mediated by the simultaneous expression of ZFNs and the delivery of heterologous donor molecules for targeting a herbicide tolerant gene in *Zea mays* plants. The procedure resulted in the insertional disruption of the IPK1 locus. The IPK1 interference was accompanied by both herbicide tolerance and alteration of the inositol phosphate profile in developing seeds.

Using ZFN, Schneider *et al*. [2] were able to insert a gene expression cassette into a target region flanked by two ZFN cutting sites. The 7 kb target site was replaced by a 4 kb unit with a selection and a visual marker. The homology directed repair event was implemented and evaluated in tobacco BY-2 cells. This proof of principle may result in the generation of transgenic crops, site-specifically modified at high frequency.

Arabidopsis plants expressing ZFNs were transformed *via* floral dip with a repair T-DNA with an incomplete protoporphyrinogen oxidase (PPO) gene. This repair T-DNA, in conjunction with the expressed ZFN, induced double-strand break of the PPO gene. The repair T-DNA harboring a PPO gene, missing the 5' coding region and containing two mutations, resulted the generation of plants harboring an insensitive PPO gene and thus resistant to the herbicide butafenacil [3].

Tandemly arrayed genes (TAG) represent almost 17% of the Arabidopsis gene repertoire. Reverse genetics of these regions is usually complex and limited. Qi *et al*. [4] engineered ZFNs to target seven genes from three TAGs regions on two Arabidopsis chromosomes. One of these targets was the RPP4 gene cluster, which contains eight resistance genes. ZFN editing experiments resulted in gene cluster deletions ranging from 2 kb to 55 kb at high frequencies in somatic cells. Moreover, large chromosomal deletions of ~9 Mb and targeted chromosome rearrangements were also accomplished.

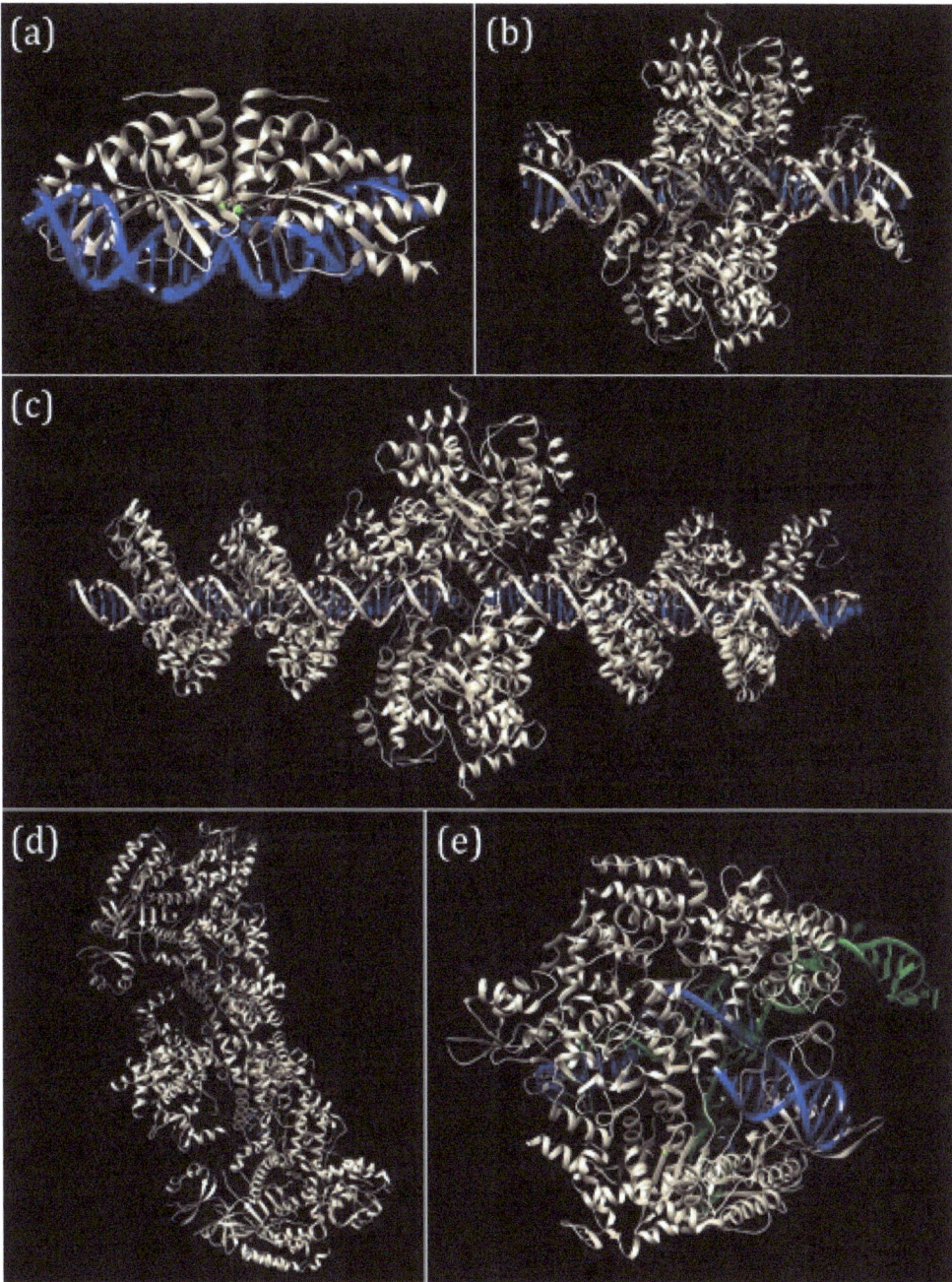

Fig. (1). 3D rendering of composite structures of diverse nucleases bound to their corresponding DNA targets: (a) meganuclease I-Cre; (b) zinc finger nuclease/FOK; (c) TALEN/FOK; (d) Cas9; (e) Cas9 + DNA/RNA. The figures were developed, compiled, and edited based on available crystal structures corresponding to PDB IDs of the protein data bank # 4AQU, 1AAY, 2FOK, and 4UN3.

TALEN

Transcription activator-like (TAL) effectors were first discovered in plant pathogenic bacteria from the genus Xanthomonas as effector proteins that are translocated into the plant host cell and interact with plant genomic DNA through a modular DNA binding domain. This domain is characterized by tandem polymorphic amino acid repeats that mediate individual binding to specific sequences of contiguous DNA nucleotides [5, 6]. These domains, associated with nucleases like FokI (TALEN, Fig. **1c**), permit site-directed mutagenesis by specific DNA targeting [7].

Zhang *et al.* [8] were able to efficiently engineer the tobacco genome by optimizing and adapting this technology. *Nicotiana tabacum* protoplasts were incubated with TALENs targeting the acetolactate synthase gene, which was successfully edited. TALEN activity was quantified with an elegant approach comprising a single-strand annealing assay in which TALEN cleavage generates a functional yellow fluorescent protein coding gene. With this assay, the authors successfully determined the optimal TALEN scaffolds for efficient target edition.

In barley, Wendt *et al.* [9] with TALEN were able to generate targeted double-strand breaks that resulted in short deletions of specific gene sequences. Firstly, they devised several TALENs and tested *in vivo* in a yeast-based annealing assay. After selection of the most efficient nucleases, barley plants were transformed with the selected TALEN and analyzed for the presence of edited sites at the target sequences, which showed short deletions in the predicted loci.

Forner *et al.* [10] generated Arabidopsis plants harboring TALEN constructs with meristem-specific promoters that efficiently targeted germline tissue, which was eventually edited. The TALEN's target was the CLV3 coding sequence; the edited transformants presented the typical CLV3 loss of function phenotype with broadened stem and distinctly misshaped club-like siliques [11].

TALENs were also designed to specifically target the endogenous early flowering vernalization FRIGIDA gene in *Brassica oleracea*. The designed TALEN was able to bind and edit the target site, and the regenerated plants presented the typical *frigida* phenotype [12].

Li *et al.* [13] designed TALENs to knock out two α(1,3)-fucosyltransferase (FucT) and two β (1,2)-xylosyltransferase (XylT) genes of *Nicotiana benthamiana* plants. The independent inhibition of these enzymes has been described as improving the capacity to produce glycoproteins devoid of plant-specific residues. The complete knockout line was employed as a platform to transiently express a recombinant antibody. The expressed antibody had Nglycans that lacked β (1,2)-xylose. The most advantageous glycoform, lacking both core α(1,3)-fucose and β (1,2)-xylose residues, was increased in the antibody from 2% when produced in the wild type line to 55% in the mutant line. These TALEN-derived *N. benthamiana* knockout lines, in combination with Agrobacterium-mediated transient expression, are presented as a tool for the production of relevant biotechnological products.

There is a natural limitation in the design of TALEN targets. The 5′-most base of the DNA sequence bound by the TALE (the N0 base) must be a thymine, and this could be a potential limitation when specific loci are intended to be edited. A major step forward for this technology was reported by Lamb *et al.* [14]. They described the use of a structure-guided library design, in parallel with live activity selections, in order to evolve novel TALEN terminal domains that would allow any N0 nucleotide. An unconstrained domain was isolated and engineered to develop a new TALEN modular domain. The newly assembled domains provided effective and specific targeting of any DNA sequence. These new TALEN terminal domains facilitate the directed editing of any target sequence.

Finally, among several TALEN-associated online toolboxes, Doyle *et al.* [15] developed a tool suitable for rapid customization of TALEN, not only in DNA targeting, but also for engineered gene regulation and as site-specific nucleases for genome editing.

Meganucleases

Meganucleases (MN, reviewed in [16]) are natural rare-cutting endonucleases or "molecular scissors". They are larger than ZFN and bind to their target DNA by numerous contacts. Moreover, their DNA binding domains are generally associated with their catalytic domain. These characteristics result in a more complex system, engineered in order to be redirected to selected target sequences. Furthermore, although they present relatively higher selectivity to their targets, their overall efficiency is lower than other nucleases, making them less versatile and ultimately less exploited.

The I-Cre MN [17] (Fig. **1a**) has been engineered to edit maize plants. An MN, comprised of a dimer of I-*Cre*I fused into a polypeptide, was devised to recognize a target sequence at the *LIGULELESS1* gene promoter. In the resulting mutated plants (3%), short deletions and insertions were found at the intended target locus.

The modifications were heritable, and the offspring presented the characteristic liguleless phenotype [18].

The yeast MN I-SceI [19, 20] was also employed to modify maize. Targeted mutagenesis by double-stranded breaks of the I-*Sce*I directed site occurred at a rate of 1%, these being the most predominant editing short deletions [21]. Although there are several interesting achievements derived from the use of this technology [16], an apparent shift toward more efficient nucleases has been observed in the literature, as illustrated by the exponential growth of reports related to ZFN, TALEN, and CRISPR/Cas.

CRISPR/Cas 9

CRISPR/Cas 9 (Fig. **1d-e**) is the most important breakthrough technology of the decade. In its few years of existence there has been an explosion of related literature that is flooding every life science field. Its widely diverse applications range from human health, to ecological drives, to crop improvement. In this chapter we will focus only on the most recent progress associated with crop breeding.

Among new optimizations of the CRISPR-based technology, it is worth mentioning a newly designed robust CRISPR/Cas9 vector system, which includes a plant codon optimized Cas9 gene for highly efficient multiplex genome editing in monocot and dicot plants. Moreover, a PCR-based procedure was developed to swiftly generate multiple sgRNA expression cassettes, to be assembled into a binary CRISPR/Cas9 vector by simple cloning into the Golden Gate ligation or Gibson Assembly platform. While utilizing this approach, the authors were able to edit several target genes in rice with a high rate of mutation. This versatile newly developed tool has vast potential for studying multiple genes in both monocot and dicot plants [22].

There has recently been a broadened interest in the identification and implementation of Cas9 orthologues. This is not only based on the possibility of developing newly efficient alternatives within the CRISPR/Cas9 platform, but is also directed toward overcoming possible intellectual property issues associated with the conflicting CRISPR patents submitted independently by the Doudna and Zhang groups. (*i.e.,* Doudna: CA2872241A1, https://patents.google.com/patent/ CA2872241A1; Zhang: US8697359, http://www.google.com/patents/ US8697 359).

As an example, Steinert *et al.* [23] have very recently optimized Cas9 orthologues from *Streptococcus thermophilus* (St1Cas9) and *Staphylococcus aureus* (SaCas9). These enzymes were applied to induce error-prone NHEJ targeted mutagenesis in Arabidopsis. Moreover, SaCas9 and SpCas9 proteins induced homologous recombination through double-strand breaks in the presence of their single-guide (sg) RNAs, which are dissimilar and species-specific. This characteristic is not trivial, because it allows the simultaneous targeting of different sequence motifs with different enzyme efficiencies in the same plant cell without interference, licensing multiple independent targeting of endogenous genes.

Also, specific advances have been described regarding the tissue-specific expression of the modifying enzymes. Wang *et al.* [24] have specifically expressed Cas9 in Arabidopsis egg cells and one cell stage embryos, inducing the generation of homozygous or biallelic T1 mutants for multiple targets with a relatively high efficiency (8.3%). This rapid and effective approach has optimized the CRISPR/Cas platform.

An analog platform has been recently presented by Mao *et al.* [25]. It also involves the design of a germline-specific Cas9 system, oriented to genic editing of male gametocytes through a SPOROCYTELESS (SPL) expression cassette. With this approach, two endogenous genes were targeted. Although the mutations were rare in T1 generation, the rate increased by 30% in the second generation. This approach thus generates mostly T2 heterozygotes, presenting a diversity of mutations in the targeted locus. This tool could be useful for the production of genotypes of interest in T2 populations, especially of intended mutations that could be lethal.

Li *et al.* [26] reported the successful application of CRISPR/Cas in Glycine max. They were able to induce targeted mutagenesis, gene integration, and gene editing in soybean. They selected two genomic sites, *DD20* and *DD43*, on chromosome 4, encoding the acetolactate synthase1 (ALS) gene, which was modified at frequencies ranging from 59% to 76%. Interestingly, they were able to specifically target only one of the four soybean *ALS* homologous genes, highlighting the high specificity of the method and the possibility of independently modifying highly identical genes of the same family.

Perhaps one of the most interesting developments, in terms of both the economic and the sustainability perspective, is the generation of virus resistant plants by directing the CRISPR/Cas system to target DNA virus genes. *Geminivirus* are plant infecting viruses of the *Begomoviridae* family. They are highly destructive, affecting several relevant crops and generating important economic losses. The resistant genotypes are highly limited, and most managing practices involve insecticide-mediated control of their insect vectors, the *Bemicia tabaci* white flies. Bisaro and Voytas [27] have reported a breakthrough control strategy of geminivirus (potentially extrapolable to any DNA virus). The *Bean yellow dwarf virus* (BeYDV) genome was targeted for destruction with a CRISPR/Cas system incorporated in transient assays employing BeYDV-based replicons. The CRISPR/Cas reagents introduced mutations within the viral genome and reduced virus copy number. Moreover, transgenic plants expressing CRISPR-guided Cas 9 targeting the virus, when challenged with BeYDV, reduced virus load and symptoms.

Furthermore, Čermák *et al.* [28] have employed geminivirus-derived replicons as a shuttle of CRISPR/Cas9 genes. Using this method they obtained frequencies tenfold higher than traditional Agrobacterium methods of DNA delivery, and were able to create heritable modifications to the *Solanum lycopersicum* genome. Moreover, the site-specific genomic editing was transmitted to progeny. Importantly, no evidence was found of persistent extra-chromosomal replicons or off-target integration of T-DNA or replicon sequences. The utilization of this technology could be extended to other geminivirus host crops, or might be emulated by virus replicons from different viral families.

There have been several attempts to apply the CRISPR/Cas system to edit essential genes that mediate the infection process of plant pathogens. For instance, Jia *et al.* [29] reported the modification of a susceptibility gene in citrus, concomitant with an alleviation of symptoms related to canker. Citrus canker is caused by *Xanthomonas citri* subspecies citri (Xcc); it is a severe disease for citrus cultivars and is responsible for significant economic losses worldwide. The authors reported the editing of the effector binding elements of the promoter of the CsLOB1 susceptibility gene. In the infection process, CsLOB1 was induced by the PthA4 effector. The PthA4 effector was unable to bind and induce the mutated CsLOB1 promoters, resulting in mild canker and alleviated infection.

Lawrenson *et al.* [30] explored the target-specificity requirements of CRISPR-guided Cas9 genome editing in *Hordeum vulgare* and *Brassica oleracea*. They were able to obtain, in the first generation, mutations in 23% and 10% of the lines, respectively. Interestingly, they targeted multiple copy genes of both plants and analyzed the potential editing of the related, non-targeted copy. In both species, they observed off-target activity in non-target gene sequences, with one mismatch between the single-guide RNA and the non-target gene sequences. These observations raise the possibility of potentially inhibiting the activity of unintended targets, or of utilizing this phenomenon to concurrently mutate a gene family by selecting a highly conserved gene sequence.

Graham and Root [31] have suggested a framework intended to improve the reliability of CRISPR/Cas editing experiments. They recommend a plethora of *in silico* tools for selecting sgRNAs to maximize the likelihood of high activity and specificity. They emphasize the use of multiple sgRNAs, and introduce an optimal number of sgRNAs per gene target, ranging from three to eight. Furthermore, they highlight the necessity of confirmation and determination of on-target efficacy and the experimental assessment of off-target effects of the utilized sgRNAs. These basic foundations should increase the success of the implementation of this important engineering tool.

There is a potential in the near future for developing fine-tuning tools to specifically switch on or off the activity of the CRISPR/Cas-guided system. This could likely be valuable when genetic modifications are intended only in exceptional conditions, for instance those of biotic or abiotic stress. A recent discovery may be implemented into technology to this end. Bondy-Denomy *et al.* [32] have identified certain proteins produced by phages that inhibit the CRISPR/Cas system. These proteins are able to block the DNA-binding activity of the CRISPR/Cas complex, consequently inhibiting the potential DNA editing by different methods. The modes of actions of these proteins range from interacting with different protein subunits, steric or non-steric modes of inhibition, or by binding to the Cas3 helicase and precluding its recruitment to the CRISPR/Cas DNA complex.

In order to diminish the eventual off-target effect, Slaymaker *et al.* [33] have developed an elegant experimental approach, devising a highly precise Cas9 enzyme. The process involves a structure-guided protein engineering of *Streptococcus pyogenes* Cas9 (SpCas9), deep sequencing of specific targets, and unbiased whole genome off-target analysis to assess Cas9-guided DNA cleavage. The resulting "enhanced specificity" SpCas9 (eSpCas9) variants present reduced off-target effects while maintaining efficient on-target cleavage.

An alternative toolbox has been developed by Lowder *et al.* [34]. It is intended to streamline and facilitate rapid and wide-scale CRISPR/Cas9-guided editing. They have developed a protocol and reagents implemented by the Golden Gate and Gateway cloning methods to efficiently assemble CRISPR/Cas9 DNA constructs for several model monocots and dicots. The availability of these platforms is instrumental to rapidly amplifying the achievements in the breeding field of hundreds of labs around the globe.

ODM

Oligo-directed mutation (ODM) consists of site-specific modification and/or target mutation (base substitution, addition, or deletion) of a gene by the delivery of chemically synthesized oligonucleotides. These oligos recognize their target by sequence identity, and induce the template-mediated substitution of target bases. The delivery of oligonucleotides is not dependable in vectors such as Agrobacterium, and may be performed by diverse methods such as transfection, particle bombardment, or electroporation [35].

ODM was developed several years ago, and it has been widely utilized in model plants and crops for functional plant genomics through direct manipulation of target genes [36 - 39].

Among several other biotechnological applications, ODN has been efficiently useful for generating the relevant trait of herbicide tolerance/resistance in rice, wheat, maize, and tobacco [40 - 44].

Intragenesis and Cisgenesis

An intragenic crop derives from the genetic modification of a plant, utilizing as transformation constructs only genetic assemblies of the same species or sexually compatible species. Intragenesis allows the generation of new-to-nature genes by the combination of functional genetic elements such as promoters, coding parts, or terminators of existing genes, rearranging them and incorporating them into plants. This new-to-nature creation differentiates this technology from cisgenics, which includes the genetic transformation of plants with only "natural" genic elements of the same or sexually-related species, without further rearrangements or generation of new combinations of functional elements [45].

Some potential projected outcomes of intragenesis include the modification of the spatial or temporal expression of endogenous genes by cognate regulatory elements, or silencing by simultaneous expression of sense and antisense regions of endogenous genes that could induce the RNAi silencing pathway and result in the specific inhibition of target genes [46].

A beneficial less stringent regulation of intragenic and/or cisgenic crops would reduce the processing costs of reaching commercialization. This could be potentially and specifically advantageous to small sized breeding companies [47], and, more importantly, to public institutions devoted to biotechnology, as the latter are usually less efficient in gathering the significant funds required to bear regulatory costs.

One of the first approaches toward the intragenesis model was described by Rommens *et al.* [48]. They implemented a species of plant-derived (P-) DNA to replace the systemically utilized Agrobacterium transfer (T-) DNA in constructs, so as to incorporate into the plant only native DNA. The P borders were complemented with a marker-free transformation derived from linking a positive selection for temporary marker gene expression to a negative selection against marker gene integration. This experimental model was applauded and highly cited. However, several technical concerns regarding the report, and dubious insights into the real nature of the P-border sequence, resulted in the retraction of the article a few years ago [49], rendering questionable the reliability and consistency of the reported results.

Cisgenesis, on the other hand, involves the incorporation of genes from crop plants themselves or from crossable species. It has the advantage over traditional breeding techniques in that the gene transfer occurs in one stet and without the heave of several hundred genes. At the same time, the use of same-species gene is an argument in favor of treating cisgenic plants as classically bred plants [50].

This technology is intended to reduce the stringent oversight of regulatory agencies, and allow the possibility of exemption from regulation [51].

While the European Commission is in the process of discussing the latest techniques, it is highly feasible that cisgenic plants, despite not having foreign DNA, could still be classified as GMOs based on the current process-based standard [52]. This ambiguity transcends the EU commission and has reignited an interesting dispute in some sectors of the scientific community [53, 54].

To illustrate the importance of this kind of debate, a paradigmatic case study can be seen in India. In a country facing vast food insecurity, with over 200 million people suffering from some lack of nourishment, agriculture-based production has a strong influence on people's access to food. There have been several pool studies oriented toward characterizing consumer acceptance of biotechnological crops [55]. In the particular case of cisgenics, Indian respondents did not report differences in attitudes toward GMO or cisgenic rice. Interestingly, respondents were eager to pay a premium for any of those rices if they presented a "no fungicide needed" trait that could have been incorporated in the crop by cisgenics or "conventional" GM. GMO crops have the potential to alleviate food insecurity problems by increasing the food supply and simultaneously decreasing production costs. This is essential in countries where food security is an issue. But in the end, perhaps the most important outcome of this pool is the report that 76% and 73% of the respondents affirmed a willingness to consume GMO and cisgenic derived foods. It is important to highlight that sometimes the most aggressive rejections of biotechnological crops are found in countries with reduced food insecurity. The questioning of new technologies potentially utilized for food production is biased toward well-fed populations. Sometimes there seems to be an interesting and ethically questionable uncertainty in some sectors, who appear to be more concerned about the potential detriment derived from what people eat, as opposed to whether they actually eat.

In large surveys conducted in the United States and the European Union, the results are in essence strongly different from those obtained in India. In general, US and European respondents reported higher rates of approval for consuming intra/cisgenic crops (ranging from 52%-81%). However, strikingly, only 14%-33% of the respondents would consume "traditional" GMOs, that is, those harboring genes from unrelated species [56 - 58].

Accelerated Breeding

Accelerated Breeding is depicted as an improvement in the breeding status of target plants by means of diverse technologies that permit the possibility of reducing the time required for genetic improvement programs [59]. This could be achieved, for instance, by the generation and utilization of GMO intermediates which have been modified to achieve early flowering, thus reducing the generation rate. After the selected traits are incorporated in the target crop, the transgene directing the short juvenile phase is segregated by outcrossing, and the obtained crop does not present any foreign DNA. Accelerated breeding is extremely valuable for woody plants where the generation periods are considerably long [60].

There have been important advances in the understanding of the flowering process in recent years. Most key floral transition regulators have been identified and characterized in model plants and several crops. These genes are not only being incorporated in breeding programs, but their regulatory circuits are also being exploited by biotechnological approaches, so as to be able to specifically switch the reproductive status of target crops. There is vast potential in the study of floral transition genes, and moreover in the discovery of variants that determine changes in the floral phase of crops. Also, the generation of markers associated to these key genes is being incorporated into breeding platforms [61].

Reverse Breeding

Reverse breeding aims at the production of parental lines for a specific heterozygous plant. This objective relies on the use of a method for reducing the genetic recombination in the selected heterozygote by eliminating the possibility of meiotic crossing over.

The male and female spores presenting combinations of non-recombinant parental chromosomes are cultured *in vitro*, where they are induced for diplodization, resulting in the generation of homozygous doubled haploid plants. Among these plants, complementary lines are selected and used to reconstitute the heterozygote, and this process can be extended in time, resulting in the desirable fixation of target heterozygous genotypes [62].

The most advanced proof of concept for reverse breeding has been reported by Wijnker *et al.* [63] and relies on the silencing of DMC1, which encodes the meiotic recombination protein DISRUPTED MEIOTIC cDNA1. Hybrids of *A. thaliana* with silenced DMC1 were not able to recombine during meiosis, and eventually parental lines could be regenerated from vigorous hybrid individuals by *in vitro* culture of gametes and diploidization, obtaining half the genome of the original hybrid. The complementing parental pairs of homozygous founder lines were used to regenerate the original hybrid by intercrossing. The prospective use of this technology in relevant crops which rely on hybrid vigor is of special interest to the seed industry.

Grafting

While grafting is a millenary practice, there has been a revival of the possibility of implementing, for commercial products, the generation of transgenic rootstocks that would eventually transmit the incorporated trait to the scions. There are diverse examples of this practice, mostly oriented toward biotic stress protection of scions by transformation of rootstock with constructs that induce an RNA interference response directed to viral pathogens. The RNAi signal is able to spread systemically, recognize an invading virus, and inhibit infection by sequence-specific cleavage of invading viral RNAs. The rationale for this tool is understandable in crops where several scion varieties are grafted into a single rootstock, for instance in the case of citrus or vine. An alternative recent example, oriented toward abiotic stress, is the graft transmission of high temperature tolerance in tomato to non-transgenic scions from transgenic rootstocks, where the fatty acid desaturase gene (LeFAD7) was RNA-silenced. The grafted scions grown under high temperature were able to cope with the adverse conditions [64].

Other Techniques

Next generation sequencing (NGS) has introduced a paradigm shift in experimental plant science. NGS is continuously redefining our vision and interpretation of plant genomes and revolutionizing marker-assisted selection by the generation of massive amounts of information, initiating the path of genomics-assisted breeding. NGS is transforming the gene discovery pipelines, launching a renaissance in the importance of wild relatives and land races as genetic resources [65]. We are experiencing only a glimpse of the potential of NGS in breeding, with advances in the analysis of big data and large-scale phenotyping tools. Sooner rather than later, these tools will be routinely employed, and genetic improvement will likely be more efficient and resourceful. However, there is a need for the development of new analytical tools to facilitate the integration of large amounts of information from diverse sources in order to inaugurate a knowledge-based crop improvement era.

The advancing field of epigenetics could be exploited in the near future as a promising alternative for crop improvement. It would be interesting to understand the potential of selection of favorable epigenetic status, while addressing the stability of the acquired states [66]. Moreover, there is encouraging evidence of the latent implementation of novel epialleles in breeding programs, and of the managing of transgene expression by means of RNA-directed DNA methylation (RdDM) [67]. RdDM derives from the specific targeting of RNA molecules to regulatory gene sequences, which results in transcriptional silencing and concomitant modification of methylation marks, thus stabilizing the silenced phenotype [68, 69]. RdDM is usually triggered by the endogenous or heterologue production of double-stranded RNA, which is eventually processed by the RNA interference cellular silencing machinery, leading in specific cases to methylation of histones and DNA sequences bound by these histones. This process does not involve the modification of the corresponding DNA sequence, nor does it generate heritable changes at the sequence level. These attributes present this research field with an interesting perspective in terms of regulatory issues. The possibility of generating valuable phenotypes without altering the genomic component of the target plant, thus eventually circumventing the regulatory process, could be highly appreciated by breeders.

Regulatory Framework of NBT

The standard for GMO regulation, and the foundation for international reference, is the Cartagena Protocol on Biosafety (CPB). The CPB states the most prevalent definition of living modified organism (GMO, LMO, also "biotechnological" crops, "transgenic" crops, *etc.*), highlighting the possession of a novel combination of genetic material is obtained by the use of "modern bio-technology". This protocol considers modern biotechnology the application of *in vitro* nucleic acid techniques, recombinant DNA, direct injection of nucleic acid into cells or organelles, and fusion of cells beyond the taxonomic family [70].

It is worth noting that there are no scientific reasons for grouping the new breeding techniques together. This artificial definition seeks to establish a contrast with traditional GMO crops, the technological standard that has been commercially available for more than 20 years [71].

In the United States, which is the largest producer of GMOs in the world, editing plants with Agrobacterium is regulated by the Animal and Plant Health Inspection Service (APHIS). This important but time-consuming and expensive process has been the emblematic path of commercially available transgenic crops. Several technical turnarounds have been devised, associated much more with trying to avoid this process, and much less with the versatility of the technique or the science behind the developed product.

A recent development portrays the essence of this particular conundrum. Lu *et al.* [72] and Woo *et al.* [73] were able to generate and assemble the CRISPR/Cas9 system outside the plant, and introduce the resulting protein complex into the target plant. This technique is able to mutate specific genes in several important crops such as tobacco, tomato, and rice, without the need for utilizing Agrobacterium for delivery. Moreover, the resulting plants do not present any foreign DNA, and the generated mutation could not be distinguished from that obtained by traditional methods or naturally occurring genetic variation. No recombinant DNA is used in this process, and the resulting edited plants could be exempted from GMO regulations. The regulatory decision taken in this instance could eventually generate a widespread use of this technology for crop improvement.

These reports raise several questions, and at the present time the US has appointed a multi-year initiative to review federal regulations on agricultural biotechnology in order to update the regulation prerequisites in the context of the new breeding techniques. This is not an isolated case, and most countries are actively discussing these technologies at present. Perhaps the first legal framework encompassing the new breeding techniques has been developed recently by the Argentinean government [74].

What it is important to highlight is that in the particular case of the CRISPR system, the delivery of purified nucleases results in much lower mutagenesis frequencies in comparison with Agrobacterium-mediated methods. As mentioned above, there is a need to discuss an important question related to the essence of scientific discovery: Are we advancing toward the generation of innovative and breakthrough technologies, or are we continuously devoting inventiveness and resources to circumventing perhaps aged legal regulations in order to generate new products?

The definition and regulation status of GMOs has been slow to arrive at an international consensus. Different criteria in diverse regions generate not only barriers for trade, which would imply mainly economic effects, but also an important impediment to reaching a general understanding in relation to the safety and science behind the generation of GMOs.

In this scenario, the new breeding techniques come forward as a means to group a range of diverse new technologies that not only advance by improving the efficiency of genome editing and plant transformation, but also, in the case of most of them, by being oriented toward exemption from the regulatory status quo.

Thus technical advancement is accompanied by an increase in the debate and a need for clarification, in the sense of either applying old regulations to new technologies or considering updating or redesigning the regulatory framework to be able to cope with a new technological era.

The European Union could be considered as process-based rather than product-based in terms of regulatory standards. As an example, the ODM technique edits sequences in their natural genomic background, does not introduce new genes into target plants, and does not involve homologous recombination. The resulting edited sequence might be considered equivalent to naturally occurring mutations, and more importantly could be untraceable and undistinguishable from natural ones. This not only raises the question of whether these techniques would fall under standard regulatory practices. It also raises the question of whether regulatory entities would be able to review these developments, something which would depend only on the good faith of the producers to present potential products as either technological developments or naturally occurring variants. In other words, according to a process-based standard, ODMs might not be considered GMOs and could be exempted from regulation [75]. The United States, through APHIS, has considered that ODM-derived plants are not GMOs [76], and an ODM-modified canola will be on the market this year. Australia, Brazil and, Argentina, though mostly using process-based standards, tend to be consistent with US standards [77].

These ambiguities should be contemplated and reinterpreted. There is no more room for gray in this discussion. An ODN-derived crop is or is not a GMO. Ambiguous or case-by-case standards tend to enlarge the regulatory definitions and may impede the implementation of economically relevant crops. The European Union also adheres to the precautionary principle, considering that if the scientific data fails to be sufficient to allow for a complete risk evaluation, the development should be stopped from distribution or withdrawn from the market. The final decision involves the analysis of so-called "sufficient scientific data", which may or may not be subject to disagreement, inconsistencies, or ambiguities.

CONCLUDING REMARKS

Several new tools are being applied to crop improvement every day, based on breakthrough versatile platforms that are increasing effectiveness and speed in the generation of new varieties of crops. The latest achievements are not only adaptations or improvements of modern techniques such as intra- and cis-genesis, accelerated breeding based on transgenics, and null-segregants, meganucleases, and zinc finger nucleases; we are also seeing the dawn of new disruptive technologies that are reshaping the paradigm of genetic improvement, as is the case with TALEN and CRISPR/Cas. Site-directed genome editing has become precise, cost-effective, versatile, and fast. These new breeding platforms are leading the way to next generation biotechnology.

CONFLICT OF INTEREST

The authors confirm that this chapter content has no conflict of interest.

ACKNOWLEDGEMENTS

Humberto J. Debat is a researcher at the Instituto Nacional de Tecnología Agropecuaria (INTA-Argentina). The author would like to thank P. Schubert for helpful discussions.

REFERENCES

[1]   Shukla VK, Doyon Y, Miller JC, *et al.* Precise genome modification in the crop species *Zea mays* using zinc-finger nucleases. Nature 2009; 459(7245): 437-41.
      [http://dx.doi.org/10.1038/nature07992] [PMID: 19404259]

[2]   Schneider K, Schiermeyer A, Dolls A, *et al.* Targeted gene exchange in plant cells mediated by a zinc finger nuclease double cut. Plant Biotechnol J 2015; 14(4): 1151-60.
      [http://dx.doi.org/10.1111/pbi.12483] [PMID: 26426390]

[3]   de Pater S, Pinas JE, Hooykaas PJ, van der Zaal BJ. ZFN-mediated gene targeting of the Arabidopsis protoporphyrinogen oxidase gene through Agrobacterium-mediated floral dip transformation. Plant Biotechnol J 2013; 11(4): 510-5.

[4]   Qi Y, Li X, Zhang Y, *et al.* Targeted deletion and inversion of tandemly arrayed genes in *Arabidopsis thaliana* using zinc finger nucleases. G3: Genes| Genomes| Genetics 2013; 3(10): 1707-15.

[5]   Boch J, Scholze H, Schornack S, *et al.* Breaking the code of DNA binding specificity of TAL-type III effectors. Science (New York, NY) Band 326, number 5959, December. 2009.
      [http://dx.doi.org/10.1126/science.1178811]

[6]   Moscou MJ, Bogdanove AJ. A simple cipher governs DNA recognition by TAL effectors. Science (New York, NY) Band 326, Number 5959, December. 1501.

[7]   Bogdanove AJ, Voytas DF. TAL effectors: customizable proteins for DNA targeting. Science 2011; 333(6051): 1843-6.
      [http://dx.doi.org/10.1126/science.1204094] [PMID: 21960622]

[8]   Zhang Y, Zhang F, Li X, *et al.* Transcription activator-like effector nucleases enable efficient plant genome engineering. Plant Physiol 2013; 161(1): 20-7.
      [http://dx.doi.org/10.1104/pp.112.205179] [PMID: 23124327]

[9]   Wendt T, Holm PB, Starker CG, *et al.* TAL effector nucleases induce mutations at a pre-selected location in the genome of primary barley transformants. Plant Mol Biol 2013; 83(3): 279-85.
      [http://dx.doi.org/10.1007/s11103-013-0078-4] [PMID: 23689819]

[10] Forner J, Pfeiffer A, Langenecker T, Manavella PA, Lohmann JU. Germline-transmitted genome editing in *Arabidopsis thaliana* Using TAL-effector-nucleases. PLoS One 2015; 10(3): e0121056.
      [http://dx.doi.org/10.1371/journal.pone.0121056] [PMID: 25822541]

[11] Müller R, Bleckmann A, Simon R. The receptor kinase CORYNE of Arabidopsis transmits the stem cell-limiting signal CLAVATA3 independently of CLAVATA1. Plant Cell 2008; 20(4): 934-46.

[http://dx.doi.org/10.1105/tpc.107.057547] [PMID: 18381924]

[12]   Sun Z, Li N, Huang G, *et al.* Site-specific gene targeting using transcription activator-like effector (TALE)-based nuclease in *Brassica oleracea.* J Integr Plant Biol 2013; 55(11): 1092-103.
[http://dx.doi.org/10.1111/jipb.12091] [PMID: 23870552]

[13]   Li J, Stoddard TJ, Demorest ZL, *et al.* Multiplexed, targeted gene editing in *Nicotiana benthamiana* for glyco-engineering and monoclonal antibody production. Plant Biotechnol J 2015; 14(2): 533-42.
[http://dx.doi.org/10.1111/pbi.12403] [PMID: 26011187]

[14]   Lamb BM, Mercer AC, Barbas CF III. Directed evolution of the TALE N-terminal domain for recognition of all 5 bases. Nucleic Acids Res 2013; 41(21): 9779-85.
[http://dx.doi.org/10.1093/nar/gkt754] [PMID: 23980031]

[15]   Doyle EL, Booher NJ, Standage DS, *et al.* TAL Effector-Nucleotide Targeter (TALE-NT) 2.0: tools for TAL effector design and target prediction. Nucleic Acids Res 2012; 40(Web Server issue): W117-22.
[PMID: 22693217]

[16]   Hafez M, Hausner G. Homing endonucleases: DNA scissors on a mission. Genome 2012; 55(8): 553-69.
[http://dx.doi.org/10.1139/g2012-049] [PMID: 22891613]

[17]   Jurica MS, Monnat RJ Jr, Stoddard BL. DNA recognition and cleavage by the LAGLIDADG homing endonuclease I-CreI. Mol Cell 1998; 2(4): 469-76.
[http://dx.doi.org/10.1016/S1097-2765(00)80146-X] [PMID: 9809068]

[18]   Gao H, Smith J, Yang M, *et al.* Heritable targeted mutagenesis in maize using a designed endonuclease. The Plant J 2010; 61(1): 176-87.

[19]   Puchta H, Dujon B, Hohn B. Two different but related mechanisms are used in plants for the repair of genomic double-strand breaks by homologous recombination. Proc Natl Acad Sci USA 1996; 93(10): 5055-60.
[http://dx.doi.org/10.1073/pnas.93.10.5055] [PMID: 8643528]

[20]   Duan X, Gimble FS, Quiocho FA. Crystal structure of PI-SceI, a homing endonuclease with protein splicing activity. Cell 1997; 89(4): 555-64.
[http://dx.doi.org/10.1016/S0092-8674(00)80237-8] [PMID: 9160747]

[21]   Yang M, Djukanovic V, Stagg J, *et al.* Targeted mutagenesis in the progeny of maize transgenic plants. Plant Mol Biol 2009; 70(6): 669-79.
[http://dx.doi.org/10.1007/s11103-009-9499-5] [PMID: 19466565]

[22]   Ma X, Zhang Q, Zhu Q, *et al.* A robust CRISPR/Cas9 system for convenient, high-efficiency multiplex genome editing in monocot and dicot plants. Mol Plant 2015; 8(8): 1274-84.
[http://dx.doi.org/10.1016/j.molp.2015.04.007] [PMID: 25917172]

[23]   Steinert J, Schiml S, Fauser F, Puchta H. Highly efficient heritable plant genome engineering using Cas9 orthologues from *Streptococcus thermophilus* and *Staphylococcus aureus.* Plant J 2015; 84(6): 1295-305.
[http://dx.doi.org/10.1111/tpj.13078] [PMID: 26576927]

[24]   Wang ZP, Xing HL, Dong L, *et al.* Egg cell-specific promoter-controlled CRISPR/Cas9 efficiently generates homozygous mutants for multiple target genes in Arabidopsis in a single generation. Genome Biol 2015; 16(1): 144.
[http://dx.doi.org/10.1186/s13059-015-0715-0] [PMID: 26193878]

[25]   Mao Y, Zhang Z, Feng Z, *et al.* Development of germ-line-specific CRISPR-Cas9 systems to improve the production of heritable gene modifications in *Arabidopsis.* Plant Biotechnol J 2015; 14(2): 519-32.
[PMID: 26360626]

[26]   Li Z, Liu ZB, Xing A, *et al.* Cas9-Guide RNA Directed Genome Editing in Soybean. Plant Physiol 2015; 169(2): 960-70.

[http://dx.doi.org/10.1104/pp.15.00783] [PMID: 26294043]

[27] Bisaro DM, Voytas DF. Conferring resistance to geminiviruses with the CRISPR–Cas prokaryotic immune system. Nature Plants 2015; 1: 15145.
[http://dx.doi.org/10.1038/nplants.2015.145]

[28] Čermák T, Baltes NJ, Čegan R, Zhang Y, Voytas DF. High-frequency, precise modification of the tomato genome. Genome Biol 2015; 16(1): 232.
[http://dx.doi.org/10.1186/s13059-015-0796-9] [PMID: 26541286]

[29] Jia H, Orbovic V, Jones JB, Wang N. Modification of the PthA4 effector binding elements in Type I CsLOB1 promoter using Cas9/sgRNA to produce transgenic Duncan grapefruit alleviating XccΔpthA4: dCsLOB1. 3 infection. Plant Biotechnol J 2015; 14(5): 1291-301.
[http://dx.doi.org/10.1111/pbi.12495] [PMID: 27071672]

[30] Lawrenson T, Shorinola O, Stacey N, *et al.* Induction of targeted, heritable mutations in barley and *Brassica oleracea* using RNA-guided Cas9 nuclease. Genome Biol 2015; 16(1): 258.
[http://dx.doi.org/10.1186/s13059-015-0826-7] [PMID: 26616834]

[31] Graham DB, Root DE. Resources for the design of CRISPR gene editing experiments. Genome Biol 2015; 16(1): 260.
[http://dx.doi.org/10.1186/s13059-015-0823-x] [PMID: 26612492]

[32] Bondy-Denomy J, Garcia B, Strum S, *et al.* Multiple mechanisms for CRISPR-Cas inhibition by anti-CRISPR proteins. Nature 2015; 526(7571): 136-9.
[http://dx.doi.org/10.1038/nature15254] [PMID: 26416740]

[33] Slaymaker IM, Gao L, Zetsche B, Scott DA, Yan WX, Zhang F. Rationally engineered Cas9 nucleases with improved specificity. Science 2015; 351(6268): 84-8.
[http://dx.doi.org/10.1126/science.aad5227] [PMID: 26628643]

[34] Lowder LG, Zhang D, Baltes NJ, *et al.* A CRISPR/Cas9 toolbox for multiplexed plant genome editing and transcriptional regulation. Plant Physiol 2015; 169(2): 971-85.
[http://dx.doi.org/10.1104/pp.15.00636] [PMID: 26297141]

[35] Sauer NJ, Mozoruk J, Miller RB, *et al.* Oligonucleotide-directed mutagenesis for precision gene editing. Plant Biotechnol J 2015.
[http://dx.doi.org/10.1111/pbi.12496] [PMID: 26503400]

[36] Hohn B, Puchta H. Gene therapy in plants. Proc Natl Acad Sci USA 1999; 96(15): 8321-3.
[http://dx.doi.org/10.1073/pnas.96.15.8321] [PMID: 10411868]

[37] Zhu T, Peterson DJ, Tagliani L, St Clair G, Baszczynski CL, Bowen B. Targeted manipulation of maize genes *in vivo* using chimeric RNA/DNA oligonucleotides. Proc Natl Acad Sci USA 1999; 96(15): 8768-73.
[http://dx.doi.org/10.1073/pnas.96.15.8768] [PMID: 10411950]

[38] Beetham PR, Kipp PB, Sawycky XL, Arntzen CJ, May GD. A tool for functional plant genomics: chimeric RNA/DNA oligonucleotides cause *in vivo* gene-specific mutations. Proc Natl Acad Sci USA 1999; 96(15): 8774-8.
[http://dx.doi.org/10.1073/pnas.96.15.8774] [PMID: 10411951]

[39] Gamper HB, Parekh H, Rice MC, Bruner M, Youkey H, Kmiec EB. The DNA strand of chimeric RNA/DNA oligonucleotides can direct gene repair/conversion activity in mammalian and plant cell-free extracts. Nucleic Acids Res 2000; 28(21): 4332-9.
[http://dx.doi.org/10.1093/nar/28.21.4332] [PMID: 11058133]

[40] Zhu T, Mettenburg K, Peterson DJ, Tagliani L, Baszczynski CL. Engineering herbicide-resistant maize using chimeric RNA/DNA oligonucleotides. Nat Biotechnol 2000; 18(5): 555-8.
[http://dx.doi.org/10.1038/75435] [PMID: 10802626]

[41] Dong C, Beetham P, Vincent K, Sharp P. Oligonucleotide-directed gene repair in wheat using a transient plasmid gene repair assay system. Plant Cell Rep 2006; 25(5): 457-65.

[http://dx.doi.org/10.1007/s00299-005-0098-x] [PMID: 16404599]

[42]   Okuzaki A, Toriyama K. Chimeric RNA/DNA oligonucleotide-directed gene targeting in rice. Plant Cell Rep 2004; 22(7): 509-12.
[http://dx.doi.org/10.1007/s00299-003-0698-2] [PMID: 14634786]

[43]   Kochevenko A, Willmitzer L. Chimeric RNA/DNA oligonucleotide-based site-specific modification of the tobacco acetolactate synthase gene. Plant Physiol 2003; 132(1): 174-84.
[http://dx.doi.org/10.1104/pp.102.016857] [PMID: 12746523]

[44]   Iida S, Terada R. Modification of endogenous natural genes by gene targeting in rice and other higher plants. Plant Mol Biol 2005; 59(1): 205-19.
[http://dx.doi.org/10.1007/s11103-005-2162-x] [PMID: 16217613]

[45]   Schouten HJ, Jacobsen E. Cisgenesis and intragenesis, sisters in innovative plant breeding. Plant Biotechnol J 2008; 6: 135-45.
[PMID: 17784907]

[46]   Molesini B, Pii Y, Pandolfini T. Fruit improvement using intragenesis and artificial microRNA. Trends Biotechnol 2012; 30(2): 80-8.
[http://dx.doi.org/10.1016/j.tibtech.2011.07.005] [PMID: 21871680]

[47]   Holme IB, Wendt T, Holm PB. Intragenesis and cisgenesis as alternatives to transgenic crop development. Plant Biotechnol J 2013; 11(4): 395-407.
[http://dx.doi.org/10.1111/pbi.12055] [PMID: 23421562]

[48]   Rommens CM, Humara JM, Ye J, *et al.* Crop improvement through modification of the plants own genome. Plant Physiol 2004; 135(1): 421-31.
[http://dx.doi.org/10.1104/pp.104.040949] [PMID: 15133156]

[49]   Rommens CM, Humara JM, Ye J, *et al.* Retractions: Crop improvement through modification of the plants own genome. Plant Physiol 2013; 161(4): 2182.
[PMID: 23549622]

[50]   Jacobsen E, Schouten HJ. Cisgenesis strongly improves introgression breeding and induced translocation breeding of plants. Trends Biotechnol 2007; 25(5): 219-23.
[http://dx.doi.org/10.1016/j.tibtech.2007.03.008] [PMID: 17383037]

[51]   Schouten HJ, Krens FA, Jacobsen E. Do cisgenic plants warrant less stringent oversight? Nat Biotechnol 2006; 24(7): 753.
[http://dx.doi.org/10.1038/nbt0706-753] [PMID: 16841052]

[52]   Kanchiswamy CN, Malnoy M, Velasco R, Kim JS, Viola R. Non-GMO genetically edited crop plants. Trends Biotechnol 2015; 33(9): 489-91.
[http://dx.doi.org/10.1016/j.tibtech.2015.04.002] [PMID: 25978870]

[53]   Schouten H. Reply to Cisgenesis as a golden mean. Nat Biotechnol 2014; 32(8): 728-8.
[http://dx.doi.org/10.1038/nbt.2981] [PMID: 25101742]

[54]   Eriksson D, Stymne S, Schjoerring JK. The slippery slope of cisgenesis. Nat Biotechnol 2014; 32(8): 727-7.
[http://dx.doi.org/10.1038/nbt.2980] [PMID: 25101741]

[55]   Shew AM, Nalley LL, Danforth DM, *et al.* Are all GMOs the same? Consumer acceptance of cisgenic rice in India. Plant Biotechnol J 2016; 14(1): 4-7.
[http://dx.doi.org/10.1111/pbi.12442] [PMID: 26242818]

[56]   Gaskell G, Allansdottir A, Allum N, *et al.* The 2010 Eurobarometer on the life sciences. Nat Biotechnol 2011; 29(2): 113-4.
[http://dx.doi.org/10.1038/nbt.1771] [PMID: 21301431]

[57]   Lusk JL, Rozan A. Consumer acceptance of ingenic foods. Biotechnol J 2006; 1(12): 1433-4.
[http://dx.doi.org/10.1002/biot.200600187] [PMID: 17124706]

[58]   Evenson RE, Santaniello V, Eds. Consumer acceptance of genetically modified foods. CABI 2004.
[http://dx.doi.org/10.1079/9780851997476.0000]

[59]   Comeau A, Caetano VR, St-Pierre CA, Haber S. Accelerated Breeding: Dream or Reality? Wheat in a Global Environment Developments in Plant Breeding 2001; 9: 671-9.
[http://dx.doi.org/10.1007/978-94-017-3674-9_90]

[60]   Flachowsky H, Hanke MV, Peil A, Strauss SH, Fladung M. A review on transgenic approaches to accelerate breeding of woody plants. Plant Breed 2009; 128(3): 217-26.
[http://dx.doi.org/10.1111/j.1439-0523.2008.01591.x]

[61]   Jung C, Müller AE. Flowering time control and applications in plant breeding. Trends Plant Sci 2009; 14(10): 563-73.
[http://dx.doi.org/10.1016/j.tplants.2009.07.005] [PMID: 19716745]

[62]   Dirks R, van Dun K, de Snoo CB, *et al.* Reverse breeding: a novel breeding approach based on engineered meiosis. Plant Biotechnol J 2009; 7(9): 837-45.
[http://dx.doi.org/10.1111/j.1467-7652.2009.00450.x] [PMID: 19811618]

[63]   Wijnker E, van Dun K, de Snoo CB, *et al.* Reverse breeding in *Arabidopsis thaliana* generates homozygous parental lines from a heterozygous plant. Nat Genet 2012; 44(4): 467-70.
[http://dx.doi.org/10.1038/ng.2203] [PMID: 22406643]

[64]   Nakamura S, Hondo K, Kawara T, *et al.* Conferring high-temperature tolerance to nontransgenic tomato scions using graft transmission of RNA silencing of the fatty acid desaturase gene. Plant Biotechnol J 2015; 14(2): 783-90.
[http://dx.doi.org/10.1111/pbi.12429] [PMID: 26132723]

[65]   Barabaschi D, Tondelli A, Desiderio F, *et al.* Next generation breeding. Plant Sci 2016; 242: 3-13.
[http://dx.doi.org/10.1016/j.plantsci.2015.07.010] [PMID: 26566820]

[66]   Springer NM. Epigenetics and crop improvement. Trends Genet 2013; 29(4): 241-7.
[http://dx.doi.org/10.1016/j.tig.2012.10.009] [PMID: 23128009]

[67]   Mahfouz MM. RNA-directed DNA methylation: mechanisms and functions. Plant Signal Behav 2010; 5(7): 806-16.
[http://dx.doi.org/10.4161/psb.5.7.11695] [PMID: 20421728]

[68]   Wassenegger M. RNA-directed DNA methylation. Plant Mol Biol 2000; 43(2-3): 203-20.
[http://dx.doi.org/10.1023/A:1006479327881] [PMID: 10999405]

[69]   Aufsatz W, Mette MF, van der Winden J, Matzke AJ, Matzke M. RNA-directed DNA methylation in Arabidopsis. Proc Natl Acad Sci USA 2002; 99 (Suppl. 4): 16499-506.
[http://dx.doi.org/10.1073/pnas.162371499] [PMID: 12169664]

[70]   Bail C, Falkner R, Marquard H, Eds. The Cartagena Protocol on Biosafety: Reconciling trade in biotechnology with environment and development. Routledge 2014.

[71]   Brookes G, Barfoot P. Key environmental impacts of global genetically modified (GM) crop use 19962011. GM Crops Food 2013; 4(2): 109-19.
[http://dx.doi.org/10.4161/gmcr.24459] [PMID: 23635915]

[72]   Luo S, Li J, Stoddard TJ, *et al.* Non-transgenic Plant Genome Editing Using Purified Sequence-Specific Nucleases. Mol Plant 2015; 8(9): 1425-7.
[http://dx.doi.org/10.1016/j.molp.2015.05.012] [PMID: 26074033]

[73]   Woo JW, Kim J, Kwon SI, *et al.* DNA-free genome editing in plants with preassembled CRISPR-Cas9 ribonucleoproteins. Nat Biotechnol 2015; 33(11): 1162-4.
[http://dx.doi.org/10.1038/nbt.3389] [PMID: 26479191]

[74]   Whelan AI, Lema MA. Regulatory framework for gene editing and other new breeding techniques (NBTs) in Argentina. GM Crops Food 2015; 6(4): 253-65.
[http://dx.doi.org/10.1080/21645698.2015.1114698] [PMID: 26552666]

[75]   Breyer D, Herman P, Brandenburger A, *et al.* Genetic modification through oligonucleotide-mediated
       mutagenesis. A GMO regulatory challenge? Environ Biosafety Res 2009; 8(2): 57-64.
       [http://dx.doi.org/10.1051/ebr/2009007] [PMID: 19833073]

[76]   Wolt JD, Wang K, Yang B. The Regulatory Status of Genome-edited Crops. Plant Biotechnol J 2015;
       14(2): 510-8.
       [http://dx.doi.org/10.1111/pbi.12444] [PMID: 26251102]

[77]   Smyth SJ, Phillips PW. Risk, regulation and biotechnology: the case of GM crops. GM Crops Food
       2014; 5(3): 170-7.
       [http://dx.doi.org/10.4161/21645698.2014.945880] [PMID: 25437235]

CHAPTER 5

Use of Plant Virus and Post-Transcriptional Gene Silencing for Plant Biotechnological Applications

Gabriela Conti[*], Andrea L. Venturuzzi, Diego Zavallo, Verónica C. Delfosse and Yamila C. Agrofoglio

Institute of Biotechnology – Center of Veterinary and Agricultural Research – IB – CICVyA – INTA, Argentina

Abstract: The current interest in green technologies has directed attention to the use of plant systems for several applications, including traditional crop plant systems used for biomass production, large-scale synthesis of a great number of recombinant proteins, and biofuels production. In this context, plant viruses are very useful instruments for plant biotechnology applications, constituting suitable tools for heterologous gene expression. Virions are particles with a complex composition, but their stability allows them to be used for the development of numerous biotechnological applications and as research tools for plant functional genomics studies. The development of infectious full-length viral clones is a strategy extensively employed as an alternative tool for introducing viruses into plants *via* inoculation with *Agrobacterium tumefaciens*. Another strategy, called RNA interference, a plant gene expression regulation mechanism based on post-transcriptional gene silencing, has extensively been employed to down-regulate the expression of endogenous transcripts and displays a number of biotechnological applications. Additionally, transgenic expression of viral proteins has been used to achieve pathogen-derived resistance, a mechanism that confers resistance to viral infections in agricultural crops. In this chapter we will discuss several strategies and methods for plant gene expression which employ plant viruses developed over the past decade.

Keywords: Agroinfiltration, Infectious clones, Plant virus, Post-transcriptional gene silencing, RNA interference, Viral expression vectors, Virus-induced gene silencing.

INTRODUCTION

Plant systems constitute advantageous systems in comparison with other conventional systems used to express recombinant proteins (mammals, bacterial or yeast cells). They offer lower production costs, and they also synthesize

[*] **Corresponding author Gabriela Conti:** Institute of Biotechnology – Center of Veterinary and Agricultural Research – IB – CICVyA – INTA, Argentina; Tel: 0054 11 64243517; E-mail: conti.gabriela@inta.gob.ar

Daniela Defavari do Nascimento, & William A. Pickering (Eds.)

proteins that are structurally and functionally equivalent to those produced by mammalian cells. They thus present the advantage of being secure and with no risks of animal pathogen contamination, and also have the possibility of easily scaling up the purification technologies [1 - 5]. Furthermore, the use of specific promoters may allow the delivery of recombinant proteins (*e.g.,* vaccines) through different plant organs such as fruits, tubers, leaves, or seeds. In this way, the cold chain required for storage and transport of purified recombinant products can be avoided, as well as administration procedures by injection, in the case of orally administrated products like vaccines [1].

On the other hand, viruses constitute very powerful tools for plant biotechnology applications. Plant viruses can be efficient vehicles for heterologous gene expression in plants, which can be used as biofactories or biofuels sources. Virions are highly complex and stable nanostructures and constitute the basis for the development of multiple biotechnological applications. Plant viruses have also been used as research tools for plant functional genomics studies (by means of the virus-induced gene silencing strategy, which targets and down-regulates specific host transcripts). RNA interference is a plant protection approach based on post-transcriptional gene silencing. It has been developed to interfere with virus infections in plants and also to down-regulate the expression of endogenous transcripts. In addition, transgenic expression of viral proteins has been employed as a means to achieve pathogen-derived resistance to viral infection. In this sense, the development of infectious full-length viral clones is a strategy extensively employed as an alternative tool for introducing virus into plants *via* inoculation with *Agrobacterium tumefaciens*. In this chapter, we will discuss several plant-virus systems used for expression of proteins and other biotechnological applications involving transgenic expression of viral proteins and induction of post-transcriptional gene silencing.

Plant Viruses are Vehicles for Heterologous Gene Expression

Stable genetic transformation protocols are required for classical exogenous and foreign protein expression systems in plants. This strategy is based on the incorporation of a candidate gene into a plant genome and the subsequent expression of active proteins. Although many such proteins are produced by this strategy, this process is highly expensive and time consuming. Plant viral vectors constitute very useful and efficient alternatives for expressing foreign proteins [6]. A virus designed and engineered to express a candidate transcript is subsequently inserted in a host plant to initiate replication and consequently to produce significant amounts of the desired protein. The rapid viral multiplication rates in plant cells result in accumulation of exogenous transcripts into the complex viral expression system. Thus, the advantages achieved with the employment of viral vectors are these: a relatively small and easy construct amenable for manipulation; the proteins can be produced in a very fast manner, achieving huge yields; the gene is never inserted into the plant genome [7]. However, the size and complexity of the inserted transcripts are factors implicated in the success of the stable expression. Additionally, there are some concerns related to the ability of modified plant viruses to spread in the environment. Overall, in spite of the aforementioned difficulties, there are numerous plant virus vectors developed and already employed in foreign protein expression plant systems. This technology is relatively more attractive to the public because of the absence of the negative connotations usually related to genetically modified plants [6].

Nowadays, two different strategies are possible for the application of plant-made biologics in the field: first generation vectors, that use complete or full versions of the viruses, and second generation vectors, composed of partial or deconstructed viral vectors. The elimination of some functions is employed to reduce or limit undesired features of the expression system. The subsequent rebuild of the virus is achieved by replacing the eliminated required functions in a genetically modified host that is able to supply the lost functions in the virus, or, alternatively, supply similar functions derived from other systems different from viruses [7]. Some advances have been made in both strategies, for example, full virus vectors are employed to produce long polypeptides directly fused to the viral capsid protein (polypeptides of 140 amino acids). Additionally, a novel viral vector generation has been developed to expose reactogenic amino acids on the viral surface with the aim of permitting easy chemical conjugation over the viral surface, with other separately produced proteins. This interesting tool is being used to synthesize new vaccines by exposing antigens attached to the surface of the virus [7].

It is worth mentioning that technical advances in the field of plant virus expression vectors have also been accompanied by important progress in the means of introducing viruses into host plants. Rather than merely infecting plants with the appropriate viral vectors, plant leaves can now be inoculated by agroinfiltration, that is, by incorporating the virus vector into *Agrobacterium tumefaciens* and infiltrating the leaf using a syringe or a vacuum treatment [8].

Different Plant Viruses used for Plant Biofactories Development

One of the major prerequisites for the construction of plant biofactories with good exogenous gene expression levels is the appropriate selection of viral vectors. There are several plant virus expression systems that have been developed to date, mainly based on positive sense RNA viruses.

For instance, *Tobacco Mosaic Virus* (TMV), an extensively studied and described plant virus, was the first one to be employed for vector development. While original versions of TMV-based vectors depended on usage of the entire viral genome, second generation vectors were deconstructed and consisted of genome sections that were important for replication; these were subcloned into a variety of plasmid constructs [9]. For example, Lindbo *et al.* [10] determined that TMV vectors lacking the capsid protein (TRBO) driving expression under the control of a 35S promoter were a very efficient agroinfection strategy. The coat protein deleted vectors proved to be excellent tools for transient expression, due to their improved ease of use and higher amounts of exogenous protein expression.

Potato Virus X (PVX) and related Potexviruses have also been constructed into expression vectors for vaccine production. For instance, the *Human Papillomavirus-16* L2 minor capsid protein has been expressed in plants as part of a fusion protein with the PVX CP [11].

It is important to point out that other Potexviruses (such as the *White Clover Mosaic Virus, Foxtail Mosaic Virus*, and *Alternanthera Mosaic Virus*) and plant DNA viruses, including *Geminivirus* and *Cauliflower Mosaic Virus* (CaMV), have also been used for foreign gene expression [9].

Overall, plant virus vectors are very useful expression systems which can be considered as more efficient biofactories than bacteria or yeast/mammalian cells for the production of high-value molecules. They are easy to manipulate, and the infection process is very efficient, thus making viral vectors a very attractive approach that could replace the use of transgenic systems in the field of exogenous plant protein expression.

Strategies for Development of Viral Infectious Clones

On some occasions, the most common mechanism for plant viral infections in plant hosts is through insect vectors that carry viral strains and permit the entry of the virus in the moment of feeding on plants. For experimental assays in plant-virus interaction studies, and also for viral vector expression approaches, the populations of insects are maintained under controlled conditions and the manipulation of the insects to achieve a correct and efficient introduction of insects into the plants is extremely complex. The use of viral infectious clones is a very interesting strategy for bypassing technical difficulties in achieving successful infections under experimental conditions.

Infectious clone production is always the first step in reverse genetic studies of DNA and RNA plant virus studies, including foreign expression approaches. The construction of infectious RNA or DNA clones of plant viruses has become a standard laboratory technique, providing an excellent tool for the study of viral gene functions and virus-host interactions. They are helpful in expression, replication, and protein function identification studies of viruses. Many of these studies are carried out by employing *in vitro* mutagenesis (deletions, insertions, substitutions) and complementation. Nevertheless, the assembly of infectious clones requires many laborious cloning and sub-cloning steps [12].

Viruses with DNA Genome

The Geminiviridae family contains species with monopartite (2.5-3 kb) and bipartite (4.8-5.6 kb, known as DNA-A and DNA-B) single-stranded circular DNA genome. The intergenic region contains a potential stem-loop structure carrying a conserved sequence (TAATATTAC) known as origin of replication (ori), where ssDNA synthesis is initiated. While the viral replicase protein completes the replication of the rolling circle, the host DNA polymerase is used for DNA replication itself [13].

The strategy formerly used to produce DNA virus infectious clones begins by developing a restriction enzyme map of the whole viral genome and the subsequent detection of the ori region. For an infectious clone, it is necessary to isolate double-stranded DNA by differential centrifugation and different subsequent precipitation steps. It is important to mention that genome replication in *planta* produces double-stranded intermediates. Thus, purified DNA is partially digested and the construct is achieved by coupling the complete genome with a duplicated ori region [14].

Nowadays, a simpler and more powerful method is described for cloning DNA genomes. It is based on a rolling circle amplification (RCA) strategy and the use of a DNA polymerase derived from a phi-29 bacteriophage. This polymerase has a unique strand displacement ability and produces a high molecular weight double-stranded DNA, facilitating the procedure for cloning the circular DNA genome of the virus. This method is very useful for the development of DNA infectious clones using RCA for genome enrichment and partial digestion. Prior to agroinoculation in plants, it is necessary to purify the dimeric molecules and then insert them in a binary vector.

This approach starts with a total DNA extraction from infected plants (Fig. **1**). Then a phi-29 DNA polymerase reaction is carried out to produce the multimeric genome concatamers. While the duplicated ori region enhances the efficiency of infection, the dimer cloning strategy requires the generation of two genomic units, which are achieved by partial digestion of the multimeric genome concatamers. Next, dimeric units are purified and cloned in a binary vector to be finally electroporated in *Agrobacterium tumefaciens* [15].

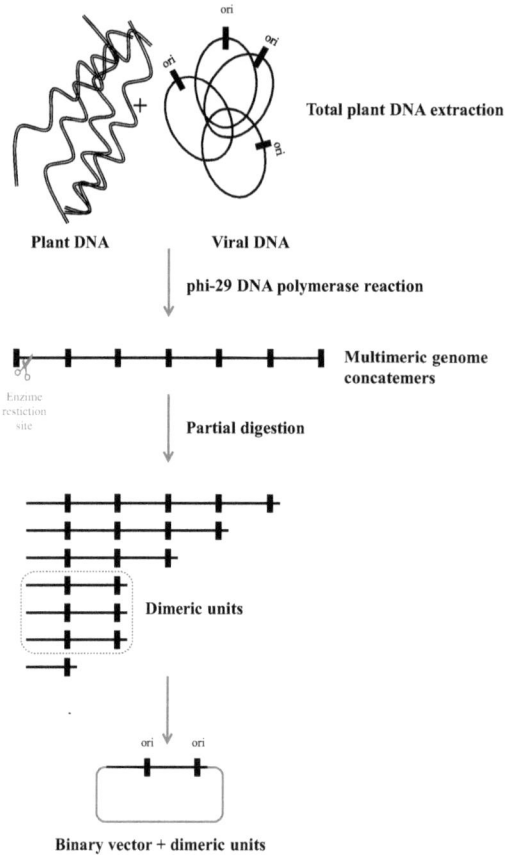

Fig. (1). Representation of the DNA genome cloning based on the rolling circle amplification method.

For bipartite viruses, both DNA components are necessary for plant infection. Therefore, bipartite virus infectious clone inoculation must be performed preparing a mixture of exactly the same amounts of component A and B in the bacterial suspensions.

It must be considered that the definition of the correct partial digestion condition to achieve the highest amount of dimeric molecules is the most laborious and limiting step in the process.

Viruses with RNA Genome

The construction of most infectious plant viral RNAs in the 1980s was achieved by *in vitro* transcription using either the T7, T3, or lambda phage promoters obtained from plasmids carrying genomic cDNAs [16]. However, many efforts were made to develop infection strategies that are more efficient and cheaper than *in vitro* transcription and manual inoculation. The agroinfection procedure is being increasingly used for inoculation of RNA virus infectious clones. Viral cDNAs are being expressed by using 35S promoter from CaMV. When the viral cDNA enters the cells, the RNA polymerase II from the host recognizes the plant promoter and initiates the synthesis of infectious RNA clones. Finally, a poly (A) tail must be added in the terminal part of cDNA in order to increase the transcript stability [17, 18].

Virus-Induced Gene Silencing (VIGS)

The term VIGS (virus-induced gene silencing) was employed to describe the mechanism by which plants can recover from viral infections [19]. It is now an extensively employed technique, based on the expression of recombinant viral vectors carrying an inserted sequence which triggers the post-transcriptional gene silencing (PTGS) of specific host genes. This technique has proved to be a potent tool for characterizing gene functions in plants, mainly after the increased availability of EST sequencing projects which constantly generate information to be checked functionally.

Functional genomic studies were formerly carried out mainly by reverse genetics, *via* Agrobacterium-mediated transformation (to design mutant knockout plants for all genes), and also by using mutagenesis approaches, either individual or in combination with tilling protocols [20 - 24].

VIGS technology provides a powerful tool for performing functional genomics studies in a faster mode, without the need to carry out time-consuming stable transformation protocols. Another advantage is that the key regulatory genes used by plants to survive are not amenable to be studied by employing mutant analyses, given their lethal pleiotropic effects in stable knockout plants. They can instead be easily analyzed by VIGS technology [25, 26], which is similar to inducible transgene expression systems. The time required to achieve systemic gene silencing is then very rapid. It is enough to silence a gene with only partial sequence information. VIGS can either be employed to silence transcripts produced by multiple copy genes (in case of polyploid plants like wheat) and also for multiple family members.

VIGS technology involves different steps. First, the engineering of viral genomes and defined host gene sequence inserts that will be targeted by gene silencing. Next, the selection of an appropriate host plant to be inoculated with the engineered virus. Subsequently, the triggering of systemic silencing of the target genes as a consequence of the antiviral endogenous defense mechanisms activated in the host plant [25 - 29]. Gene silencing begins when DNA and RNA viruses produce double-stranded RNA (dsRNA) [30]. This dsRNA is sliced to produce siRNAs by Dicer-Like Protein 4 (DCL4) [31]. The process can be extended by amplification with endogenous RNA-dependent RNA polymerase 1 (RDR1), RDR2, and RDR6, which employ the single-strand RNA as a template to generate more dsRNAs [32]. Antiviral response by target-directed gene silencing is produced when the siRNAs from the virus associate with Agronaute 1 (AGO1) or AGO2 [33], constituting RISC complexes that guide the specific RNA degradation or translational inhibition through sequence homology recognition [31, 34, 35] (Fig. **2**).

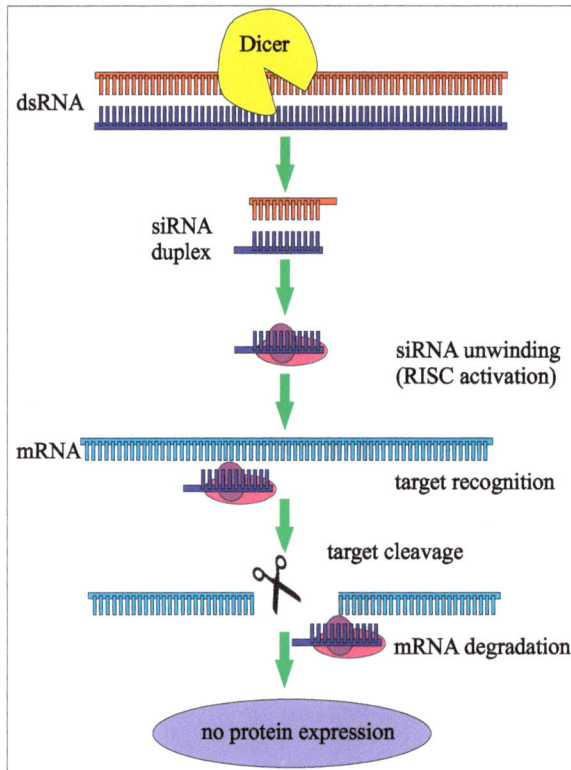

Fig. (2). Basic scheme for post-transcriptional gene silencing process.

The VIGS mechanism can be easily induced in *Nicotiana benthamiana* plants, as they are highly susceptible to viral infections enabling efficient gene silencing. *Tobacco Mosaic Virus* (TMV) was also the first viral strain to be used as a vector for inducing RNA silencing. Successful silencing of *phytoene desaturase transcript* (*pds*) was achieved by recombinant TMV transcripts carrying the *in vitro* generated *pds* sequence which was subsequently inoculated into *N. benthamiana* plants [36]. Over the years, the technology has been developed and improved to become more applicable.

The versatile recombinant VIGS vector system based on *Tobacco Rattle Virus* (TRV) is extensively used nowadays. It is composed of a bipartite RNA genome with a first segment (RNA1) containing the viral RNA-dependent RNA polymerase (RdRp) under the control of CaMV 35S promoter, then a movement protein (MP), an additional 16-kDa protein and a ribozyme. The second segment (RNA2) carries a multiple cloning site located downstream of the capsid protein (CP) controlled by CaMV 35S promoter, and a ribozyme. Each segment is inserted into the T-DNA cassettes flanked by left and right T-DNA borders, and then cloned into binary vectors which are inserted into *Agrobacterium tumefaciens*.

Agrobacterium independent cultures (one carrying the RNA1 segment, and the other with RNA2 containing the target partial sequence to be silenced) are mixed in 1:1 proportions in an appropriate buffer solution, and then used to agroinfiltrate plant leaves (usually *N. benthamiana*). Infected cells synthesize high amounts of full viral particles that produce systemic spread of the virus from the local tissues (site of infiltration) to upper systemic leaves during plant growth, triggering the gene silencing defense mechanism. This activation suppresses viral replication and, as a side effect, silences the expression of the target endogenous gene. Between three and four weeks after agroinfiltration, the phenotypes produced by silencing the target gene can be observed (Fig. **3**) [36].

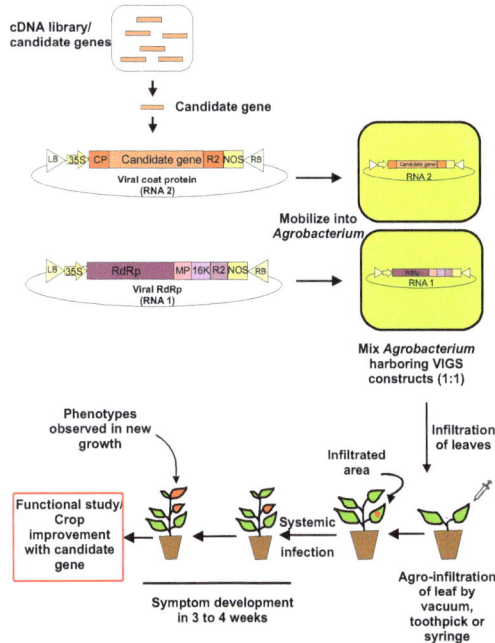

Fig. (3). Scheme representing TRV-based vector strategy employed for VIGS in plants. First, the identification of a candidate gene is required, for example from a cDNA library. The gene is next cloned into the RNA2 vector, usually by control of 35S promoter. RNA1 strain (containing genes for RdRP, MP, *etc.*) is inserted into an additional T-DNA expression system. Both binary vectors (RNA1 and RNA2) are independently inserted into *Agrobacterium thumefaciens*. The resulting recombinant Agrobacterium cultures are mixed and agroinfiltrated in *N. benthamiana*. Approximately two weeks after agroinfiltration of both VIGS vectors, a systemic infection with appearance of specific symptoms can be observed. Between two and three weeks after agroinfiltration, the phenotypes produced by endogenous candidate gene silencing can be observed (adapted from Godge *et al.* [36]).

At present, the VIGS system has been extensively used in more than 30 plant species, among them angiosperms. This strategy is useful for providing rapid characterization of genes involved in different steps of plant growth and development [26, 37 - 40], disease control and resistance [41], symbiosis [42], resistance to nematode infections [43], resistance to insects [44], stresses related to nutrient uptake [45, 46], and tolerance to abiotic stresses [26, 47].

RNA Silencing of Endogenous Transcripts in Plants

RNA silencing, or the so-called RNA interference pathway, is a plant protection mechanism based on post-transcriptional gene silencing, developed to interfere with virus infections in plants and knock down the expression of genes.

RNA silencing is a very well conserved gene expression regulation mechanism in eukaryotic cells, through which the levels of selected transcripts are limited by transcriptional suppression, called transcriptional gene silencing (TGS), or by mRNA degradation by post-transcriptional gene silencing (PTGS).

Gene silencing was initially described in plants in the 1990s, when a research group made an assay to express exogenous transgenes into petunias with the aim of increasing CHALCONE SYNTHASE activity (an enzyme implicated in specific petunia pigment production) by inducing transcript accumulation [48, 49]. Surprisingly, pigmentation was not increased, but a variegated color of petal phenotype was observed, even with total pigmentation loss in a few cases. The latter suggested that the transgene activity was lost and that the endogenous transcript expression was being inhibited. A similar phenomenon was discovered in fungus, named quelling [50], and in animals, called RNA interference (RNAi).

In nature, RNA silencing or RNAi refers to a mechanism involved not only in cellular defense against pathogens, but also for retrotransposon regulation and post-transcriptional gene expression regulation. However, it can also be used as an individual specific gene silencing tool by producing knockout plants, either by transgenesis strategy or by means of VIGS technology, as described in detail above.

In both cases, the degradation of RNA is triggered by double-stranded RNA molecules (dsRNA), which are further cleaved into 21nt duplexes. These duplexes are then loaded into a complex of enzymes that induces the mRNA degradation, leading consequently to the inhibition or suppression of target gene expression.

The RNA interference mechanism is usually produced in four basic steps: (1) First, the RNA (either double-stranded RNA or micro RNA primary transcript) is sliced to generate double-stranded short interfering RNAs (siRNA) by the RNase III enzymes (called dicer-like proteins, DCLs); (2) Next, the small dsRNA are loaded into endonuclease enzymes (Argonaute proteins, AGOs), which catalyze the unwinding of the siRNA duplex; (3) The guide siRNA strand loaded into AGO protein is subsequently incorporated into an RNA-induced silencing complex (RISC), releasing the other strand. (4) Finally, the guide strand is employed by RISC complex to find and induce the endonucleolytic cleavage or translational inhibition of the complementary mRNA target sequence.

Another efficient tool extensively applied for functional genomic studies is insertional mutagenesis. It is very useful for functional characterization of genes and ultimately for biotechnological purposes. However, this approach has limitations, such as the presence of redundant genes, the lethality of some knock-out genes, the presence of non-tagged mutants, and the lack of precision when trying to target the mutating sequence onto a specific site [51].

RNAi is a tool that largely overcomes these problems through the possibility of engineering a self-complementary hairpin RNA (hpRNAs) molecule, capable of triggering the endogenous machinery and suppressing the expression of a specific transcript. Moreover, the extent and timing of the gene silencing process can be precisely controlled by the use of hairpin RNAs under inducible promoters [52].

RNAi technology has potential to be used in numerous plant species, mainly to improve crops and enhance nutritional components, but also for pathogen defense.

Wang *et al.* [53] employed this technique to develop new varieties of barley plants with improved tolerance and resistance to *Barley Yellow Dwarf Virus* (BYDV). They showed that while the control barley plants became infected, the engineered plants with RNAi technology were resistant to viral infection. The *Banana Bract Mosaic Virus* (BbrMV) infects banana plants, destroying production in Southeast Asia and India [54]. The design of an RNAi vector aimed to silence the coat protein (CP) region of the virus was able to generate a banana variety resistant to BbrMV. A large number of other different plant species, including monocots and dicots, have also been transformed to acquire virus resistance to very harmful viruses [55, 56].

Recently, the RNAi method is being applied to modulate different targets or traits related to metabolic engineering of plants.

It has shown to be a promising tool for the development of tomato (*Solanum lycopersicum*) fruits with increased amounts of flavonoids and carotenoid contents (for example the tomato fruit's typical red color, provided by high lycopene and carotenoid contents), both of which are highly beneficial to human health [57]. RNAi-mediated suppression of *De-Ethiolated 1* (DET1) expression is responsible for repressing several light-mediated pathways. It has been demonstrated that under fruit-specific promoters the carotenoid and flavonoid levels in tomato fruits can be modulated without producing impacts on plant growth and fruit quality traits [57]. The combination of RNAi technology with tissue-specific promoters is a highly efficient strategy that provides profitable traits unlikely to be obtained by conventional breeding [58].

Cellulosic biomass can potentially contribute to supplying the increasing demand for liquid fuels. However, extensive land use and inefficient extraction processes constitute other difficulties for the deployment of large-scale biomass-to-biofuel technologies [59]. RNAi technology could be useful for developing more efficient and less harmful ways of producing cellulosic biomass as alternative sources of energy.

A key agronomic trait that impacts the production of biofuels from ligno-cellulosic biomass is the lignin content of feedstock. One example is the case of sugarcane, an elite crop for production of bioethanol. The ethanol obtained from sugarcane through fermentation of sucrose can easily be extracted from stem internodes. However, the vast amount of ligno-cellulosic residues stored in the plant cell walls of leaves can be exploited and thus support the ethanol output from sugarcane.

Recently, a group of researchers down-regulated the caffeic acid O-methyltransferase (COMT) transcript by means of RNAi technology, reducing the lignin content by 13.7% [60]. This reduction resulted in an increased yield of fermentable glucose up to 34% after pretreatment, which demonstrated that a moderate reduction in lignin can reduce the recalcitrance of sugarcane biomass [60].

Another example of reduced lignin content through RNAi was found in switchgrass (*Panicum virgatum*) by the down-regulation of 4-coumarate: coenzyme, a ligase (Pv4CL1) involved in monolignol biosynthesis [61]. Significant increments of cellulose hydrolysis efficiency is achieved when the lignin biomass becomes reduced.

Alternative sources of energy different from fossil fuels are an urgent global demand, and the use of biofuels provides a potential solution. Conversion of lignocellulosic biomass, which is both abundant and renewable, appears to be a promising alternative, even though the cellulases and the pretreatment processes are very expensive. Genetic approaches such as RNA interference are promising techniques for solving this problem, by reducing the need for pretreatment processes through lignin modification.

Transgenic Plants for Expression of Viral Proteins

Host plants expressing viral proteins display a considerable number of useful applications. First of all, this system is a simplified strategy for studying plant virus interaction, because it has the potential to discriminate the impact of each individual viral protein on virus replication [62]. The ectopic expression of a P6 viral transcript from *Mosaic Cauliflower Virus* (CaMV) in *Arabidopsis thaliana*, a host species, demonstrated that this protein is capable of producing phenotypic alterations similar to infection symptoms, such as late flowering and mild chlorosis [63, 64]. Transgenic expression of the small subunit of the *Beet Western Yellow Virus* (BMYV) capsid protein P74 produced unusual patterns of PTGS when *Nicotiana benthamiana* plants were infected with this virus or other related BMYW [65]. In a similar assay, the TMV126-kDa protein expression coupled to green fluorescent protein (GFP) produced increases of susceptibility to a variety of viruses in *Nicotiana tabacum* [66].

Pathogen-derived resistance is another important biotechnological application of viral protein expression. In 1985, Sanford and Johnston proposed a method for interfering with the development of pathogenic processes through the use of pathogen-derived genes, thereby conferring disease resistance on the host plant [67]. The expression of certain viral genes in host plants upsets the normal balance of components and interferes with the replication process. This viral control strategy is called pathogen-derived resistance, and consists in the generation of transgenic plants expressing genomic sequences of the pathogen to be controlled, thereby causing reductions in viral accumulation and replication. Two main modes of conferring pathogen-derived resistance were developed. The first one, by transgenic expression of TMV capsid protein (CP), was developed by Roger Beachy's group in order to control a virus in tobacco plants. Later, it was found that even by expressing translated versions of CP or other viral sequences, resistance response could be reached [68, 69]. This mechanism was finally elucidated as an RNA-mediated resistance, being a consequence of post-transcriptional gene silencing.

Since 1986, the expression of viral capsid proteins has been widely used to confer resistance to viral infections. This mechanism, called capsid protein-mediated resistance (CP-MR), was used to confer resistance to *Alfalfa Mosaic Virus* (AIMV) [70 - 72], *Cucumber Mosaic Virus* (CMV) [73], *Potato Virus X* (PVX) [74], and *Tobacco Streak Virus* (TSV) [71], among others. Despite being widely studied, the molecular mechanisms governing CP-MR are not yet fully understood. It has been shown that the molecular bases of each resistance type are different. In the case of TMV, CP-MR is based on interference with the viral genome desencapsidation when the virus enters the cell. Given this mechanism, the resistance can be overcome when plants are inoculated with viral RNAs or partially desencapsidated virions (pretreated with pH 8.0 solution) [75]. Furthermore, it is already known that the degree of resistance is dependent on the inoculum level, thus at higher inoculum levels the resistance can be broken. Additionally, the level of transgene expression positively correlates with CP-mediated resistance [76], and this positive correlation is also reinforced with greater ability to self-assemble CP [77]. For example, when the amino acid 42 (threonine) of TMV CP is mutated to tryptophan (the new mutated version is called CP^{T42W}), the amount of 20S CP aggregates is higher and more stable [78, 79], resulting in resistance increases when CP^{T42W} is transgenically expressed in tobacco plants [78 - 80]. Moreover, some degree of resistance has also been observed when plants were inoculated with infective viral RNAs [78] and cell to cell viral movement was partially restricted [81]. Moreover, the mutant CP^{T42W} reduces the formation of viral replication complexes (VRC) [82] and, just as with the CP wild type version, it confers delayed infections with unrelated viruses such as PVX [83].

For all these reasons it has been proposed that the CP has a regulatory role that is not essential for viral replication, and which allows normal infective process development [83]. Subsequently, Bendahmane *et al.* [77] found that transgenic expression of CP, both wild type and mutated version, is capable of interfering with viral CP functions such as control of infection, virus replication, and movement.

CONCLUDING REMARKS

From all the cases mentioned in this book chapter, it is evident that plant viruses show an extremely significant potential for use as biotechnological tools. Current studies are focused on the discovery of new viruses with differential properties, or on the improvement of the already developed vectors for refining actual tools and for new applications. It is a great challenge for researchers to exploit this technology in diverse areas of agriculture such as vaccine production or the biofuels industry.

CONFLICT OF INTEREST

The authors confirm that this chapter content has no conflict of interest.

ACKNOWLEDGEMENTS

G.C. is a career research assistant, D.Z. has a postdoctoral position, and A.L.V, V.C.D., and Y.C.A. are PhD students, all at the Consejo Nacional de Investigaciones Científicas y Técnicas (CONICET-Argentina).

REFERENCES

[1]   Lico C, Desiderio A, Banchieri S, Benvenuto E. Plants as biofactories: Production of pharmaceutical recombinant proteins 2003.

[2]   Hellwig S, Drossard J, Twyman RM, Fischer R. Plant cell cultures for the production of recombinant proteins. Nat Biotechnol 2004; 22(11): 1415-22.
[http://dx.doi.org/10.1038/nbt1027] [PMID: 15529167]

[3]   Liénard D, Sourrouille C, Gomord V, Faye L. Pharming and transgenic plants. Biotechnol Annu Rev 2007; 13: 115-47.
[http://dx.doi.org/10.1016/S1387-2656(07)13006-4] [PMID: 17875476]

[4]   Lico C, Chen Q, Santi L. Viral vectors for production of recombinant proteins in plants. J Cell Physiol 2008; 216(2): 366-77.
[http://dx.doi.org/10.1002/jcp.21423] [PMID: 18330886]

[5]   Sourrouille C, Marshall B, Liénard D, Faye L. From neanderthal to nanobiotech: From plant potions to pharming with plant fact 2009.

[6]   Cañizares MC, Nicholson L, Lomonossoff GP. Use of viral vectors for vaccine production in plants. Immunol Cell Biol 2005; 83(3): 263-70.
[http://dx.doi.org/10.1111/j.1440-1711.2005.01339.x] [PMID: 15877604]

[7]   Gleba Y, Marillonnet S, Klimyuk V. Engineering viral expression vectors for plants: the full virus and the deconstructed virus strategies. Curr Opin Plant Biol 2004; 7(2): 182-8.
[http://dx.doi.org/10.1016/j.pbi.2004.01.003] [PMID: 15003219]

[8]   Leuzinger K, Dent M, Hurtado J, *et al.* Efficient agroinfiltration of plants for high-level transient expression of recombinant proteins. J Vis Exp 2013; (77): 77.
[PMID: 23913006]

[9]   Hefferon K. Plant virus expression vector development: new perspectives. Biomed Res Int 2014; 13: 785382.
[http://dx.doi.org/10.1155/2014/785382]

[10] Lindbo JA. TRBO: a high-efficiency tobacco mosaic virus RNA-based overexpression vector. Plant Physiol 2007; 145(4): 1232-40.
[http://dx.doi.org/10.1104/pp.107.106377] [PMID: 17720752]

[11] Cerovska N, Hoffmeisterova H, Moravec T, *et al.* Transient expression of *Human papillomavirus* type 16 L2 epitope fused to N- and C-terminus of coat protein of *Potato virus X* in plants. J Biosci 2012; 37(1): 125-33.
[http://dx.doi.org/10.1007/s12038-011-9177-z] [PMID: 22357210]

[12] Nagyová A, Subr Z. Infectious full-length clones of plant viruses and their use for construction of viral vectors. Acta Virol 2007; 51(4): 223-37.
[PMID: 18197730]

[13]   Varsani A, Navas-Castillo J, Moriones E, *et al.* Establishment of three new genera in the family *Geminiviridae*: *Becurtovirus, Eragrovirus* and *Turncurtovirus.* Arch Virol 2014; 159(8): 2193-203. [http://dx.doi.org/10.1007/s00705-014-2050-2] [PMID: 24658781]

[14]   Nagata T, Inoue-Nagata AK. Simplified methods for the construction of RNA and DNA virus infectious clones. Methods Mol Biol 2015; 1236: 241-54. [http://dx.doi.org/10.1007/978-1-4939-1743-3_18] [PMID: 25287508]

[15]   Ferreira P de T, Lemos TO, Nagata T, Inoue-Nagata AK. One-step cloning approach for construction of agroinfectious *begomovirus* clones. J Virol Methods 2008; 147(2): 351-4. [http://dx.doi.org/10.1016/j.jviromet.2007.10.001] [PMID: 18022703]

[16]   van Emmelo J, Ameloot P, Fiers W. Expression in plants of the cloned satellite tobacco necrosis virus genome and of derived insertion mutants. Virology 1987; 157(2): 480-7. [http://dx.doi.org/10.1016/0042-6822(87)90290-X] [PMID: 18644559]

[17]   Leiser RM, Ziegler-Graff V, Reutenauer A, *et al.* Agroinfection as an alternative to insects for infecting plants with *beet western yellows luteovirus.* Proc Natl Acad Sci USA 1992; 89(19): 9136-40. [http://dx.doi.org/10.1073/pnas.89.19.9136] [PMID: 1409615]

[18]   Liu L, Lomonossoff G. Agroinfection as a rapid method for propagating *Cowpea mosaic virus*-based constructs. J Virol Methods 2002; 105(2): 343-8. [http://dx.doi.org/10.1016/S0166-0934(02)00121-0] [PMID: 12270666]

[19]   Van Kammen A. Virus-induced gene silencing in infected and transgenic plants. Trends Plant Sci 1997; 2: 409-11. [http://dx.doi.org/10.1016/S1360-1385(97)01128-X]

[20]   Xin Z, Wang ML, Barkley NA, *et al.* Applying genotyping (TILLING) and phenotyping analyses to elucidate gene function in a chemically induced sorghum mutant population. BMC Plant Biol 2008; 8: 103. [http://dx.doi.org/10.1186/1471-2229-8-103] [PMID: 18854043]

[21]   Dong C, Vincent K, Sharp P. Simultaneous mutation detection of three homoeologous genes in wheat by High Resolution Melting analysis and Mutation Surveyor. BMC Plant Biol 2009; 9: 143. [http://dx.doi.org/10.1186/1471-2229-9-143] [PMID: 19958559]

[22]   Gady AL, Hermans FW, Van de Wal MH, van Loo EN, Visser RG, Bachem CW. Implementation of two high through-put techniques in a novel application: detecting point mutations in large EMS mutated plant populations. Plant Methods 2009; 5: 13. [http://dx.doi.org/10.1186/1746-4811-5-13] [PMID: 19811648]

[23]   Perry J, Brachmann A, Welham T, *et al.* TILLING in Lotus japonicus identified large allelic series for symbiosis genes and revealed a bias in functionally defective ethyl methanesulfonate alleles toward glycine replacements. Plant Physiol 2009; 151(3): 1281-91. [http://dx.doi.org/10.1104/pp.109.142190] [PMID: 19641028]

[24]   Chawade A, Sikora P, Bräutigam M, *et al.* Development and characterization of an oat TILLING-population and identification of mutations in lignin and beta-glucan biosynthesis genes. BMC Plant Biol 2010; 10: 86. [http://dx.doi.org/10.1186/1471-2229-10-86] [PMID: 20459868]

[25]   Burch-Smith TM, Anderson JC, Martin GB, Dinesh-Kumar SP. Applications and advantages of virus-induced gene silencing for gene function studies in plants. Plant J 2004; 39(5): 734-46. [http://dx.doi.org/10.1111/j.1365-313X.2004.02158.x] [PMID: 15315635]

[26]   Senthil-Kumar M, Rame Gowdaa H, Ramanna Hemaa B, *et al.* Virus-induced gene silencing and its applications. CAB Rev PAVSNNR 2008; 3: 1-18.

[27]   Becker A, Lange M. VIGSgenomics goes functional. Trends Plant Sci 2010; 15(1): 1-4. [http://dx.doi.org/10.1016/j.tplants.2009.09.002] [PMID: 19819180]

[28]   Lu R, Martin-Hernandez AM, Peart JR, Malcuit I, Baulcombe DC. Virus-induced gene silencing in plants. Methods 2003; 30(4): 296-303.
[http://dx.doi.org/10.1016/S1046-2023(03)00037-9] [PMID: 12828943]

[29]   Dinesh-Kumar SP, Anandalakshmi R, Marathe R, Schiff M, Liu Y. Virus-induced gene silencing. In: Grotewold E, Ed. Plant Functional Genomics. Humana Press 2003; pp. 287-93.
[http://dx.doi.org/10.1385/1-59259-413-1:287]

[30]   Ding S-W, Voinnet O. Antiviral immunity directed by small RNAs. Cell 2007; 130(3): 413-26.
[http://dx.doi.org/10.1016/j.cell.2007.07.039] [PMID: 17693253]

[31]   Llave C. Virus-derived small interfering RNAs at the core of plant-virus interactions. Trends Plant Sci 2010; 15(12): 701-7.
[http://dx.doi.org/10.1016/j.tplants.2010.09.001] [PMID: 20926332]

[32]   Donaire L, Barajas D, Martínez-García B, Martínez-Priego L, Pagán I, Llave C. Structural and genetic requirements for the biogenesis of *tobacco rattle virus*-derived small interfering RNAs. J Virol 2008; 82(11): 5167-77.
[http://dx.doi.org/10.1128/JVI.00272-08] [PMID: 18353962]

[33]   Carbonell A, Carrington JC. Antiviral roles of plant ARGONAUTES. Curr Opin Plant Biol 2015; 27: 111-7.
[http://dx.doi.org/10.1016/j.pbi.2015.06.013] [PMID: 26190744]

[34]   Ku M, Koche RP, Rheinbay E, *et al.* Genomewide Analysis of PRC1 and PRC2 Occupancy Identifies Two Classes of Bivalent Domains. PLoS Genet 2008;4(10):e1000242.
[http://dx.doi.org/10.1371/journal.pgen.1000242]

[35]   Brodersen P, Sakvarelidze-Achard L, Bruun-Rasmussen M, *et al.* Widespread translational inhibition by plant miRNAs and siRNAs. Science 2008; 320(5880): 1185-90.
[http://dx.doi.org/10.1126/science.1159151] [PMID: 18483398]

[36]   Godge MR, Purkayastha A, Dasgupta I, Kumar PP. Virus-induced gene silencing for functional analysis of selected genes. Plant Cell Rep 2008; 27(2): 209-19.
[http://dx.doi.org/10.1007/s00299-007-0460-2] [PMID: 17938933]

[37]   Yang Y, Wu Y, Pirrello J, *et al.* Silencing Sl-EBF1 and Sl-EBF2 expression causes constitutive ethylene response phenotype, accelerated plant senescence, and fruit ripening in tomato. J Exp Bot 2010; 61(3): 697-708.
[http://dx.doi.org/10.1093/jxb/erp332] [PMID: 19903730]

[38]   Zhu X, Pattathil S, Mazumder K, *et al.* Virus-induced gene silencing offers a functional genomics platform for studying plant cell wall formation. Mol Plant 2010; 3(5): 818-33.
[http://dx.doi.org/10.1093/mp/ssq023] [PMID: 20522525]

[39]   Liu Y, Schiff M, Dinesh-Kumar SP. Virus-induced gene silencing in tomato. Plant J 2002; 31(6): 777-86.
[http://dx.doi.org/10.1046/j.1365-313X.2002.01394.x] [PMID: 12220268]

[40]   Quadrana L, Rodriguez MC, López M, *et al.* Coupling virus-induced gene silencing to exogenous green fluorescence protein expression provides a highly efficient system for functional genomics in Arabidopsis and across all stages of tomato fruit development. Plant Physiol 2011; 156(3): 1278-91.
[http://dx.doi.org/10.1104/pp.111.177345] [PMID: 21531899]

[41]   van der Linde K, Kastner C, Kumlehn J, Kahmann R, Doehlemann G. Systemic virus-induced gene silencing allows functional characterization of maize genes during biotrophic interaction with *Ustilago maydis*. New Phytol 2011; 189(2): 471-83.
[http://dx.doi.org/10.1111/j.1469-8137.2010.03474.x] [PMID: 21039559]

[42]   Grønlund M, Olsen A, Johansen EI, Jakobsen I. Protocol: using virus-induced gene silencing to study the arbuscular mycorrhizal symbiosis in *Pisum sativum*. Plant Methods 2010; 6: 28.
[http://dx.doi.org/10.1186/1746-4811-6-28] [PMID: 21156044]

[43]   Mao ZC, *et al.* The new CaSn gene belonging to the snakin family induces resistance against root-knot nematode infection in pepper. Phytoparasitica 2011; 39: 151-64.
[http://dx.doi.org/10.1007/s12600-011-0149-5]

[44]   Mantelin S, Peng HC, Li B, Atamian HS, Takken FL, Kaloshian I. The receptor-like kinase SlSERK1 is required for Mi-1-mediated resistance to potato aphids in tomato. Plant J 2011; 67(3): 459-71.
[http://dx.doi.org/10.1111/j.1365-313X.2011.04609.x] [PMID: 21481032]

[45]   Pacak A, Geisler K, Jørgensen B, *et al.* Investigations of *barley stripe mosaic virus* as a gene silencing vector in barley roots and in *Brachypodium distachyon* and oat. Plant Methods 2010; 6: 26.
[http://dx.doi.org/10.1186/1746-4811-6-26] [PMID: 21118486]

[46]   Xia Z, Su X, Wu J, Wu K, Zhang H. Molecular cloning and functional characterization of a putative sulfite oxidase (SO) ortholog from *Nicotiana benthamiana*. Mol Biol Rep 2011; 39(3): 2429-37.
[http://dx.doi.org/10.1007/s11033-011-0993-x] [PMID: 21667106]

[47]   George GM, van der Merwe MJ, Nunes-Nesi A, *et al.* Virus-induced gene silencing of plastidial soluble inorganic pyrophosphatase impairs essential leaf anabolic pathways and reduces drought stress tolerance in *Nicotiana benthamiana*. Plant Physiol 2010; 154(1): 55-66.
[http://dx.doi.org/10.1104/pp.110.157776] [PMID: 20605913]

[48]   Agrawal N, Dasaradhi PV, Mohmmed A, Malhotra P, Bhatnagar RK, Mukherjee SK. RNA interference: biology, mechanism, and applications. Microbiol Mol Biol Rev 2003; 67(4): 657-85.
[http://dx.doi.org/10.1128/MMBR.67.4.657-685.2003] [PMID: 14665679]

[49]   Napoli C, Lemieux C, Jorgensen R. Introduction of chimeric chalcone synthase gene into *Petunia* results in reversible cosuppression of homologous genes in trans. Plant Cell 1990; 2(4): 279-89.
[http://dx.doi.org/10.1105/tpc.2.4.279] [PMID: 12354959]

[50]   Romano N, Macino G. Quelling: transient inactivation of gene expression in *Neurospora crassa* by transformation with homologous sequences. Mol Microbiol 1992; 6(22): 3343-53.
[http://dx.doi.org/10.1111/j.1365-2958.1992.tb02202.x] [PMID: 1484489]

[51]   Matthew L. RNAi for plant functional genomics. Comp Funct Genomics 2004; 5(3): 240-4.
[http://dx.doi.org/10.1002/cfg.396] [PMID: 18629158]

[52]   Chen S, Hofius D, Sonnewald U, Börnke F. Temporal and spatial control of gene silencing in transgenic plants by inducible expression of double-stranded RNA. Plant J 2003; 36(5): 731-40.
[http://dx.doi.org/10.1046/j.1365-313X.2003.01914.x] [PMID: 14617073]

[53]   Wang MB, Abbott DC, Waterhouse PM. A single copy of a virus-derived transgene encoding hairpin RNA gives immunity to *barley yellow dwarf virus*. Mol Plant Pathol 2000; 1(6): 347-56.
[http://dx.doi.org/10.1046/j.1364-3703.2000.00038.x] [PMID: 20572982]

[54]   Rodoni BC, Dale JL, Harding RM. Characterization and expression of the coat protein-coding region of banana bract mosaic potyvirus, development of diagnostic assays and detection of the virus in banana plants from five countries in southeast Asia. Arch Virol 1999; 144(9): 1725-37.
[http://dx.doi.org/10.1007/s007050050700] [PMID: 10542022]

[55]   Reyes CA, De Francesco A, Peña EJ, *et al.* Resistance to *Citrus psorosis virus* in transgenic sweet orange plants is triggered by coat protein-RNA silencing. J Biotechnol 2011; 151(1): 151-8.
[http://dx.doi.org/10.1016/j.jbiotec.2010.11.007] [PMID: 21084056]

[56]   Ibrahim AB, Aragão FJ. RNAi-mediated resistance to viruses in genetically engineered plants. Methods Mol Biol 2015; 1287: 81-92.
[http://dx.doi.org/10.1007/978-1-4939-2453-0_5] [PMID: 25740357]

[57]   Davuluri GR, van Tuinen A, Fraser PD, *et al.* Fruit-specific RNAi-mediated suppression of DET1 enhances carotenoid and flavonoid content in tomatoes. Nat Biotechnol 2005; 23(7): 890-5.
[http://dx.doi.org/10.1038/nbt1108] [PMID: 15951803]

[58]   Tang G, Galili G, Zhuang X. RNAi and microRNA: breakthrough technologies for the improvement of plant nutritional value and metabolic engineering. Metabolomics 2007; 3: 357-69.
[http://dx.doi.org/10.1007/s11306-007-0073-3]

[59]   Rubin EM. Genomics of cellulosic biofuels. Nature 2008; 454(7206): 841-5.
[http://dx.doi.org/10.1038/nature07190] [PMID: 18704079]

[60]   Jung JH, Fouad WM, Vermerris W, Gallo M, Altpeter F. RNAi suppression of lignin biosynthesis in sugarcane reduces recalcitrance for biofuel production from lignocellulosic biomass. Plant Biotechnol J 2012; 10(9): 1067-76.
[http://dx.doi.org/10.1111/j.1467-7652.2012.00734.x] [PMID: 22924974]

[61]   Xu B, Escamilla-Treviño LL, Sathitsuksanoh N, *et al.* Silencing of 4-coumarate:coenzyme A ligase in switchgrass leads to reduced lignin content and improved fermentable sugar yields for biofuel production. New Phytol 2011; 192(3): 611-25.

[62]   Maule A, Leh V, Lederer C. The dialogue between viruses and hosts in compatible interactions. Curr Opin Plant Biol 2002; 5(4): 279-84.
[http://dx.doi.org/10.1016/S1369-5266(02)00272-8] [PMID: 12179959]

[63]   Zijlstra C, Schärer-Hernández N, Gal S, Hohn T. *Arabidopsis thaliana* expressing the *cauliflower mosaic virus* ORF VI transgene has a late flowering phenotype. Virus Genes 1996; 13(1): 5-17.
[http://dx.doi.org/10.1007/BF00576974] [PMID: 8938975]

[64]   Cecchini E, Gong Z, Geri C, Covey SN, Milner JJ. Transgenic *Arabidopsis* lines expressing gene VI from *cauliflower mosaic virus* variants exhibit a range of symptom-like phenotypes and accumulate inclusion bodies. Mol Plant Microbe Interact 1997; 10(9): 1094-101.
[http://dx.doi.org/10.1094/MPMI.1997.10.9.1094] [PMID: 9390424]

[65]   Brault V, Pfeffer S, Erdinger M, Mutterer J, Ziegler-Graff V. Virus-induced gene silencing in transgenic plants expressing the minor capsid protein of *Beet western yellows virus.* Mol Plant Microbe Interact 2002; 15(8): 799-807.
[http://dx.doi.org/10.1094/MPMI.2002.15.8.799] [PMID: 12182337]

[66]   Harries PA, Palanichelvam K, Bhat S, Nelson RS. *Tobacco mosaic virus* 126-kDa protein increases the susceptibility of *Nicotiana tabacum* to other viruses and its dosage affects virus-induced gene silencing. Mol Plant Microbe Interact 2008; 21(12): 1539-48.
[http://dx.doi.org/10.1094/MPMI-21-12-1539] [PMID: 18986250]

[67]   Sanford JC, Johnston SA. The concept of parasite derived resistance. J Theor Biol 1985; 113: 395-405.
[http://dx.doi.org/10.1016/S0022-5193(85)80234-4]

[68]   Powell PA, Stark DM, Sanders PR, Beachy RN. Protection against *tobacco mosaic virus* in transgenic plants that express *tobacco mosaic virus* antisense RNA. Proc Natl Acad Sci USA 1989; 86(18): 6949-52.
[http://dx.doi.org/10.1073/pnas.86.18.6949] [PMID: 2476807]

[69]   Braun CJ, Hemenway CL. Expression of Amino-Terminal Portions or Full-Length Viral Replicase Genes in Transgenic Plants Confers Resistance to *Potato Virus X* Infection. Plant Cell 1992; 4(6): 735-44.
[http://dx.doi.org/10.1105/tpc.4.6.735] [PMID: 12297660]

[70]   Loesch-Fries LS, Merlo D, Zinnen T, *et al.* Expression of *alfalfa mosaic virus* RNA 4 in transgenic plants confers virus resistance. EMBO J 1987; 6(7): 1845-51.
[PMID: 16453779]

[71]   van Dun CM, Bol JF, Van Vloten-Doting L. Expression of *alfalfa mosaic virus* and *tobacco rattle virus* coat protein genes in transgenic tobacco plants. Virology 1987; 159(2): 299-305.
[http://dx.doi.org/10.1016/0042-6822(87)90467-3] [PMID: 18644569]

[72] van Dun CM, Overduin B, van Vloten-Doting L, Bol JF. Transgenic tobacco expressing *tobacco streak virus* or mutated *alfalfa mosaic virus* coat protein does not cross-protect against *alfalfa mosaic virus* infection. Virology 1988; 164(2): 383-9.
[http://dx.doi.org/10.1016/0042-6822(88)90551-X] [PMID: 3369086]

[73] Cuozzo M, *et al.* Viral protection in transgenic tobacco plants expressing the cucumber mosaic virus coat protein or its antisense RNA. Biotechnology 1988; 6: 549-57.
[http://dx.doi.org/10.1038/nbt0588-549]

[74] Hemenway C, Fang RX, Kaniewski WK, Chua NH, Tumer NE. Analysis of the mechanism of protection in transgenic plants expressing the *potato virus X* coat protein or its antisense RNA. EMBO J 1988; 7(5): 1273-80.
[PMID: 16453840]

[75] Register JC III, Beachy RN. Resistance to TMV in transgenic plants results from interference with an early event in infection. Virology 1988; 166(2): 524-32.
[http://dx.doi.org/10.1016/0042-6822(88)90523-5] [PMID: 3176344]

[76] Powell JM, Roxas D, Lambourne J, Hoefs S. Comparison of the feeding value of local browse species ILCA Annual Report. Addis Ababa, Ethiopia: ILCA 1989.

[77] Bendahmane M, Chen I, Asurmendi S, Bazzini AA, Szecsi J, Beachy RN. Coat protein-mediated resistance to TMV infection of *Nicotiana tabacum* involves multiple modes of interference by coat protein. Virology 2007; 366(1): 107-16.
[http://dx.doi.org/10.1016/j.virol.2007.03.052] [PMID: 17499327]

[78] Bendahmane M, Fitchen JH, Zhang G, Beachy RN. Studies of coat protein-mediated resistance to *tobacco mosaic tobamovirus*: correlation between assembly of mutant coat proteins and resistance. J Virol 1997; 71(10): 7942-50.
[PMID: 9311885]

[79] Asurmendi S, Berg RH, Smith TJ, Bendahmane M, Beachy RN. Aggregation of TMV CP plays a role in CP functions and in coat-protein-mediated resistance. Virology 2007; 366(1): 98-106.
[http://dx.doi.org/10.1016/j.virol.2007.03.014] [PMID: 17493658]

[80] Bazzini AA, Hopp HE, Beachy RN, Asurmendi S. Infection and coaccumulation of *tobacco mosaic virus* proteins alter microRNA levels, correlating with symptom and plant development. Proc Natl Acad Sci USA 2007; 104(29): 12157-62.
[http://dx.doi.org/10.1073/pnas.0705114104] [PMID: 17615233]

[81] Bendahmane M, Szecsi J, Chen I, Berg RH, Beachy RN. Characterization of mutant *tobacco mosaic virus* coat protein that interferes with virus cell-to-cell movement. Proc Natl Acad Sci USA 2002; 99(6): 3645-50.
[http://dx.doi.org/10.1073/pnas.062041499] [PMID: 11891326]

[82] Asurmendi S, Berg RH, Koo JC, Beachy RN. Coat protein regulates formation of replication complexes during *tobacco mosaic virus* infection. Proc Natl Acad Sci USA 2004; 101(5): 1415-20.
[http://dx.doi.org/10.1073/pnas.0307778101] [PMID: 14745003]

[83] Bazzini AA, Asurmendi S, Hopp HE, Beachy RN. *Tobacco mosaic virus* (TMV) and *potato virus X* (PVX) coat proteins confer heterologous interference to PVX and TMV infection, respectively. J Gen Virol 2006; 87(Pt 4): 1005-12.
[http://dx.doi.org/10.1099/vir.0.81396-0] [PMID: 16528051]

CHAPTER 6

Proteomics for Bioenergy Production

Fernanda Salvato[1,*], **Bruna Marques dos Santos**[2], **Marília Gabriela de Santana Costa**[2] and **Edwin Antônio Gutierrez Rodriguez**[3]

[1] *University of Campinas, Department of Plant Biology, Institute of Biology, Campinas, Brazil*

[2] *São Paulo State University, Department of Technology, Jaboticabal, Brazil*

[3] *São Paulo State University, Department of Plant Production, Jaboticabal, Brazil*

Abstract: This chapter reviews the importance and application of proteomics tools for studying bioenergy crops and microorganisms employed in sugar fermentation and/or degradation of cell wall compounds. The large volume of information from genome sequencing projects has allowed the development of many new platforms for profiling each step of the genetic information flow from DNA to protein to many molecular interactions involved in phenotype determination. In this context, proteomics is a comprehensive package of tools dedicated to identifying and characterizing protein expression in different biological systems. In this article, the application of proteomics in bioenergy production from feedstock is summarized, citing studies associated with the reduction of lignocellulose biomass recalcitrance and the discovery of more efficient enzymes for cell wall disruption and of potent microorganisms for sugar fermentation.

Keywords: Bioenergy, Biofuel, Biomass, Enzymes, Feedstock, Lignocellulose, Lignin, Maize, Mass spectrometry, Microorganisms, Oil, Proteomics, Proteome, Renewable, Saccharification, Subcellular, Sugarcane, Sorghum, Trees, 2D gel electrophoresis.

INTRODUCTION

Pressures arising from climate change and the future lack of fossil fuel are impacting the direction of the modern agriculture. Associated with an expanding population and decreased land areas for crop production, one of the major concerns in agriculture today is the need for sustainable energy sources. This panorama has challenged agriculture production and innovation, especially in the field of bioenergy supply, to find a way to a sustainable future.

---

[*] **Corresponding author Fernanda Salvato:** University of Campinas, Department of Plant Biology, Institute of Biology, Campinas, Brazil; Tel:+55 19 3429 4475; E-mail: fersalvato@gmail.com

Daniela Defavari do Nascimento, & William A. Pickering (Eds.)

Bioenergy refers to the energy produced from biological sources, mainly plants and photosynthetic algae [1]. There are two types of bioenergy sources: the primary one employs crops; the other source is the lignocellulosic residue, discarded during food or wood production [2]. A large variety of biomass feedstocks can be used to generate bioenergy (including biofuels) and bioproducts (lignocellulose derivatives). The generation of biofuels can be achieved in different ways, such as the utilization of the sugar content in biomass *via* fermentation to produce ethanol (first generation ethanol/biodiesel). Biofuel generation can also directly originate from the whole lignocellulosic biomass (second generation ethanol), from lipids extracted from algae and oil crops, or from the use of syngas generated from the gasification of biomass [1]. Thus many crops are now seen as promising bioenergy feedstock candidates in addition to their nutritional property as food. Examples of conventional (sugar/starch) feedstock are sugarcane, sugar beet, sweet sorghum, and corn starch. On the other hand, crop residues left in the field, such as sugarcane straw, corn stover, and eucalyptus bark, can serve as biomass for second generation bioenergy production. In addition, species with high-yielding biomass and broad climatic/soil adaptation (switchgrass, miscanthus, energy cane) also serve as sources for cellulosic ethanol production.

In this context, bioenergy and biomass research have become the apple of the investor's and the researcher's eye. Currently, advancements in biotechnology, including the "omics" technologies (genomics, transcriptomics, proteomics, and metabolomics), can facilitate the identification of key genes and proteins associated with lignocellulose biosynthesis, as well as those involved in the processes of biomass degradation, biomass fermentation, and plant adaptation in adverse environments. The focus of this chapter is on pointing out the role of proteomics and its major contributions to the bioenergy sector.

THE VALUE OF PROTEOMICS FOR BIOENERGY PRODUCTION

Historically, plant genetics and breeding have led to significant improvements in agriculture production, through selection of desired phenotypes related to higher tolerance to abiotic and biotic stresses, better composition, and production efficiency. Advances at the molecular level have launched plant genetics into a new level of knowledge. Since the discovery of the DNA molecular structure and the elaboration of the central dogma of molecular biology, our understanding in genetics has evolved continuously, first with the development of marker-assisted selection allowing the association of molecular markers with agronomic traits of interest, and then with numerous DNA sequencing projects initiating the genomics era. Subsequent large-scale analyses of transcripts (transcriptomics), proteins (proteomics), and metabolites (metabolomics) associated with specific environment effects have demonstrated the dynamic response of genes in determining different phenotypes. Advancements in "omics" technologies have thus contributed to the understanding of the complex dynamics and non-linearity of biological systems [3], in disagreement in some aspects with what was stated initially by the central dogma of molecular biology. The linear flux of genetic information from DNA > RNA > proteins, suffers from the influence of diverse factors and processes that alter the genetic information translated into a specific phenotype. To cite only a few of these interfering processes: RNA editing, alternative splicing, protein-protein interactions, post-translational modifications, and non-coding RNAs. It is clear that phenotype definition is not as simple as earlier thought.

Considering the modern thinking of molecular biology, the proteome, that is, the entire set of proteins expressed by a genome under certain conditions and at a specific time [4], represents the key player in biochemical processes, being closer to the phenotype than DNA markers [5]. For this reason, proteome investigation may provide great insights into the modulation of biochemical processes.

Protein abundance, protein-protein interactions, PTMs, subcellular localization, and protein turnover, are important protein properties that should be explored for understanding the dynamics of biological processes [6]. Since the year 2000, plant proteomics has made great progress, as evidenced by the increasing number of publications. Progress has been made in multiple bases of proteomics: i) new technology, with the development of highly sensitive and accurate mass spectrometers; ii) new algorithms for confident protein assignments; iii) targeted protein approaches; iv) quantitative proteomics approaches for relative and absolute quantification; v) subcellular proteomics; and vi) the development of enrichment techniques for isolation and mapping of PTMs. All this progress has been applied to the study of different issues associated with biodiversity, nutrition, crop improvement, safety, and energy sustainability [7, 8].

In the current bioenergetics world scenario, the study of energy crops and microorganisms has also gained the attention of the proteomics field. Among crops, sugarcane, sorghum, and maize have been featured as high-yield energy crops. The main research focus is on the conversion of the lignocellulose biomass and/or saccharide plant content into simple sugars or ethanol. The conversions are highly dependent on microorganisms or enzymatic reactions. Thus intense research is being conducted on the production of high biomass and sugar yield throughout crop breeding, and on the isolation of enzymes capable of efficient conversions. Plant proteomics has contributed a lot to the field, but it really is only beginning to do so with regard to bioenergy crops. Below we discuss the progress made, and the applications of proteomics to important bioenergy crops and microorganisms.

Sorghum

Sorghum is an herbaceous grass of tropical origin, which is very tolerant to arid and saline conditions. It can be classified into four different groups according to the type of product that can result: grain sorghum, forage sorghum, high-tonnage sorghum, and sweet sorghum. As a C4 crop it is characterized by its high photosynthetic efficiency resulting in high biomass yield, which is a prerequisite for biofuel production. Compared to other crops, it also has high water and nutrient use efficiencies [9]. As compared to sugarcane and maize, little has been done toward the genetic and molecular understanding of sorghum's relevant traits for biofuel production, despite all of its agronomic advantages. However, the genome sequencing of sorghum paved the way for new discoveries in this field and has established sorghum as a model system for studying other bioenergy crops such as sugarcane and Miscanthus, which carry very complex genomes [10].

The pioneering work in sorghum proteomics came with the establishment of 2D profiles of cellular and secreted proteins from suspension cells [11]. This work was complemented by the identification of these proteins by MALDI-TOF-TOF [12]. Since then, a few other studies have come out related to salt stress. Researchers have identified salinity-stress-responsive proteins in leaves of sorghum seedlings [13], the majority being involved in photosynthesis. In the same field of study, but employing sorghum plants in hydroponic solution, 2D-PAGE protein profiles were subsequently identified by MALDI/TOF-TOF [14, 15]. These studies detected differently expressed proteins related to energy metabolism, signal transduction]n, ribosome maturation, and ROS-scavenging. More recently, a comparative study of *Sorghum bicolor* in response to drought stress showed that proteins associated with the energy balance, metabolism, and chaperons were the most salient differences between drought-tolerant and sensitive genotypes [16]. Thinking about understanding the turnover of sorghum biomass, *Aspergilus nidulans* was grown on sorghum stover during a time-course [17]. Extracellular proteins were detected, such as hemicellulases, cellulases, polygalacturonases, chitinases, esterases, and lipases. The study revealed that during one day of *A. nidulans* growth, a large variety of enzymes were secreted to degrade the great majority of polysaccharides and lipids of sorghum stover.

Sugarcane

Sugarcane is a very versatile crop, which serves as the best raw material for sugar production and as a clean source for energy production. The stalk juice is used for sugar and ethanol production. Sugarcane also provides the lignocellulosic residue (bagasse) which is employed to produce heat and electricity. Modern cultivars of sugarcane were generated by crosses between species from the genus Saccharum, with *S. spontaneum* and *S. officinarum* being the main contributors, the former with vigorous growth and the latter with high sucrose content in its stalks [18]. With the increasing need for alternative sources of renewable energy and the possibility of the utilization of sugarcane biomass, the exploitation of the fiber side of the plant brought to light the old concept of "energy cane management system" defended by Alexander in Puerto Rico in the 1980s, but rejected by society and industry at that time [19, cited in 20]. Thus, the term "energy cane" was adopted to refer to cultivars that resulted from breeding programs intended to enhance biomass yield instead of sugar content.

Sugarcane has a high polyploidy genome, and its complete genome sequence is not available. Large collections of expressed sequence tags (ESTs) and cDNA microarray data, which have contributed to genetic breeding and biotechnology development [21, 22], have become publicly available. In the field of proteomics, a few studies have recently been published, keeping the field largely open for exploration. The first work in sugarcane proteomics involved the generation of 2D gel maps of stalk and leaf proteins [23]. Another study [24] later detected leaf proteins associated with osmotic stress, using the same technique of 2D gels. Proteins strongly induced in the treatment were identified as Rubisco small subunit, ATP synthase delta chain, and isoflavone reductase-like protein.

Pacheco and collaborators [25], also evaluating stress response, showed differentially delayed root proteome response to salinity stress. They observed that in the tolerant genotype, proteins associated with carbohydrate and energy metabolism, ROS detoxification, protein protection, and membrane stabilization and growth were induced after two hours of stress. The response of the same class of proteins in the sensitive genotype was observed only after 72 hours.

In a more targeted study, Cesarino and co-workers [26] characterized a number of class III peroxidases expressed during stem development. Five class III peroxidase isoforms were identified in 2D gels, suggesting post-translational modifications. Later, using sugarcane suspension cells, isolate peroxidases (putatively related to lignification, as they can oxidize syringaldazine, which is an analogue of the monolignol sinapyl alcohol) were identified [27].

Maize

Corn (*Zea mays* L.) originated in Central America and is considered a fundamental food for human and animal consumption. In addition, it has been featured in various industrial products such as biofuels [28, 29]. The main producer of ethanol from corn is the US, and grains are the feedstock most used in its production [29]. With the expansion of the ethanol industry, the demand for corn kernels and production costs has increased [30, 31]. Corn stover is an affordable and promising agricultural waste for bioethanol production, but the hydrolysis of lignocellulosic biomasses into fermentable sugars is the challenging step [31]. Studies have been developed with regard to improving the efficiency of biofuel conversion from biomass. Recently, maize lines were evaluated for cell wall properties that can influence recalcitrance and determine the success of lignocellulose conversion [32].

Studies on corn proteome are reviewed in [33]. Aiming for greater efficiency in the production of biofuels, proteomics has been mainly employed in the identification of more efficient enzymes for biomass conversion from corn stover. In this context, the approach based on LC-MS/MS data-dependent acquisition was employed to develop a synergistic enzyme cocktail that could maximize the saccharification of pretreated corn stover, while reducing enzyme usage [34]. Recently, a mixture of proteins from different microorganisms in pre-treated corn stover was analyzed using SDS-PAGE and chromatographic separation coupled to mass spectrometry. The results showed that better combinations of biomass-degrading enzymes can be established [31].

In corn kernels, the first proteomics study related to the growth and development of endosperm cells, where starch (the substrate used for bioethanol production) is synthesized and stored, was performed using iTRAQ labeling and LC-MS/MS, comparing the proteome of a mutant and the wild type. This study was the first to characterize proteins in cell layers during maize endosperm growth and development. Proteins associated with carbohydrate metabolism and cell homeostasis were differently accumulated among layers analyzed [35]. A comparative approach of two proteomic strategies, 2D DIGE and iTRAQ labeling, was conducted to evaluate the effects of a Myb transcription factor in the accumulation of pericarp pigment. Proteins associated with glycolysis, protein synthesis, defense responses, and flavonoid and lignin biosynthesis have changed their abundance in the maize mutant for Unstable factor for orange 1-1 (Ufo1-1), which regulates the epigenetic control of Myb transcription factor gene. Based on these results, it was suggested that lignin composition could be modified by alterations in the flavonoid metabolism [36].

Trees

Trees and shrubs can be used as feedstock for biofuel production, through oil extraction from plants like palm, castor bean, and *Jatropha* for biodiesel production, or by pretreatment, hydrolysis, and fermentation of sugars from woody plants to produce ethanol. The genres Pinus, Picea, Populus, Eucalyptus, Hevea, and Teca are the main woody species that compose planted forests [36], and their lignocellulosic biomass may be used as feedstock for producing "second generation" biofuels [37, 38].

A limiting factor in the conversion of biomass is cell wall resistance, and as a consequence the reduction of lignin content has also been the subject of several studies in forest species [39]. Cell wall biosynthesis is very important for wood plants, with potential use of the raw material for conversion into biofuels. Several proteins related to the biosynthesis of non-cellulosic polysaccharides are reported, with subcelullar location in the Golgi apparatus. In this context, xylem proteins of Pinus radiata located in this organelle were analyzed by proteomics using free-flow electrophoresis (FFE) and LC-MS/MS analysis. Enriched Golgi fractions were abundant in cytoskeleton proteins (actin and tubulins) suggesting an important role of the structural proteins during compression wood development [40].

In Populus, approximately 6000 proteins were detected in developing xylem, using subcellular fractionation to separate nuclear proteins with the aim of identifying low-abundance DNA-regulatory proteins. Transcription factors and chromatin associated proteins were detected using this method [41]. In another study [42], transcription factors and regulatory enzymes related to secondary cell wall biosynthesis in differentiating xylem of *Populus trichocarpa* were identified. In addition to the proteomics discovery-driven approach, a targeted approach was employed to absolutely quantify fourteen cellulosic proteins *via* selected reaction monitoring (SRM), contributing to a more comprehensive and quantitative understanding of cellulose biosynthesis [42].

The production of biodiesel from inedible oils has been stimulated, as it does not compete with food production. Jatropha (*Jatropha curcas* L.) is a woody plant with the potential to be applied in the production of biofuels due to the high oil content in its seed [43, 44]. Using techniques of proteomics such as 2D and LC-MS/MS analysis, twenty-eight differentially abundant proteins were detected between embryo and endosperm tissues. The results showed that in dry mature seeds of *Jatropha curcas*, proteins related to seed germination were detected in endosperm and embryo, indicating the early presence of these proteins in the dry mature seed [44].

The palm (*Elaeis guineensis* L.) is used commonly as cooking oil and has gradually become used for biofuel [45], because the accumulation of oil in the mesocarp can reach 90% [46]. Using iTRAQ labeling 8-plex and 2D LC-MS/MS, the mesocarp proteome of fruits classified as high and low performance in the production of oil were analyzed, with the identification of key temporal changes that contribute to the production of oil during the fruit maturation. Proteins related to sucrose metabolism, glycolysis, pentose phosphate pathway, fatty acid metabolism, and oxidative phosphorylation were detected and were found differentially accumulated, indicating an increase in carbon flux and high-energy molecules (ATP and NADH), which are required for lipid biosynthesis [46]. In a recent study on the leaf proteome of jatoba (*Hymenaea* L.) subjected to heat stress, gas chromatography and mass spectrometry were used to analyze volatile products, and the 2D-DIGE technique followed by mass spectrometry was used to detect differentially regulated proteins. The production of sesquiterpene hydro-carbons was reported, indicating that the jatoba could potentially be used in the future for biodiesel production [47].

Subcellular Proteomics Applied to Bioenergy

The combination of subcellular isolation and mass spectrometry is known as subcellular proteomics, and it is an important approach for the identification proteins of specific cellular compartments. It is highly dependent on biochemical techniques for its performance [48]. Examples of subcellular proteomics studies include the analysis of chloroplasts [49], the nuclear envelope [50], the Golgi complex proteins [51], phagosome [52], mitochondria [53], and the cell wall [54]. Each of these studies has led to the identification of associations between proteins and organelles that have never been described previously.

The reduction of proteome complexity caused by subcellular fractionation is the great advantage of this approach, which enables deeper proteome coverage and the identification of low abundance proteins. Subcellular proteomes can therefore help in the understanding of important metabolic pathways and of how to redistribute them in plants, enabling specific chemistries and accumulation of valuable products when coupled to targeted metabolic engineering [55].

The first step to success in subcellular proteomics experiments is the quality and purity of samples. Protein contaminants or cell debris can introduce artifact in the results, with the erroneous allocation of specific proteins to organelles and also quantification errors. Large numbers of protocols for plant organelle isolation have been established. They have been used mainly for high-throughput shotgun proteomic studies of chloroplasts [56] and mitochondria [57, 58], because these organelles usually result in more purified samples. These studies employ density gradients for the purification process. However, losses of about 50% of the material can occur [59], suggesting that it is important to adapt the existing protocols to each species.

The chloroplast is an important organelle responsible for different metabolic pathways such as photosynthesis and fatty acid biosynthesis, and therefore it has been extensively studied by proteomics in a variety of plants, such as *Arabidopsis thaliana* [60], wheat [49], maize [61], and the pea [62]. Available data from some of these proteome projects are accessible *via* databases such as PPDB (http://ppdb.tc.cornell.edu/), AT_Chloro (http://at-chloro.prabi.fr/at_chloro/), and plprot (http://www.plprot.ethz.ch/).

Among the important crops for bioenergy production, maize is a common target of subcellular proteomics. In the past ten years, many quantitative proteomics studies on maize chloroplasts have been conducted. These investigations have been able to provide new findings into the regulation of differentiation and biogenesis of chloroplasts, stress response, and understanding C4 photosynthetic machinery [63 - 66]. However, the photosynthetic machinery is too complex to be interpreted based only on the quantitative proteome profile. For this reason, other proteomics studies should focus on the analysis of large-scale protein modifications and interactions in order to elucidate protein networks in maize photosynthesis [61].

The assessment of the complete chloroplast genome of *Jatropha curcas* [67] allows the conduction of functional genomics assays between model plants and biofuel crops. Pinheiro and coworkers [68] presented the first in-depth plastid proteome analysis isolated from the endosperm of *J. curcas* seeds. Functional categorization showed that proteins related to amino acid metabolism comprised the most abundant category. Other functional categories showed significant representation in the proteome, such as the carbohydrate, energy, and lipid metabolism. This study contributed to outline a general overview of the biochemical pathways of fatty acid deposition and secondary metabolites in this species, which has a especial appeal as a potential source for biodiesel production [69].

The cell wall is an important biomass material, and improvements in cell wall composition can promote the availability of large quantity of high-quality sugar skeletons for the production of bioethanol. Cell wall proteomic studies are thus expected to provide important information about the regulatory mechanisms in the determination of the quality and quantity of its components [70].

Wei and coworkers [71] showed the importance of cell wall proteomics using the model plant *A. thaliana*. In this study, the great majority of the detected glycoside hydrolases are xylan- or hemicellulose-modifying enzymes. These will probably have an impact on cellulose accessibility, which is an important factor for enzymatic hydrolysis of the plant cell wall in biofuel production.

Recently, interest in plant Golgi apparatus has emerged as the hemicelluloses and pectins are synthesized and modified in these organelles. Understanding the processes carried out in the Golgi apparatus is very important for the elucidating of cell wall deposition, cellular growth, and development [51, 71].

The proteomic characterization of high-purity Golgi membranes from plants was performed by Parsons and coworkers [51] using *A. thaliana*. A set of 371 proteins was identified with a significant proportion of matrix polysaccharide biosynthesis associated- proteins. Using transient fluorescent markers, thirteen new proteins assigned to the Golgi complex were used to validate the purity of the proteome. This proteomic study was of great importance for the cell wall research field, which with the advent of biofuels has been demanding a rapid expansion in basic knowledge [51].

The plant endomembrane system comprises of membranes from the endoplasmic reticulum, the Golgi apparatus, and the plasma membrane. Functional information from their proteomes will be needed to explore the plant biomass for the development of cost-effective biofuels [72].

Lao and coworkers [73] characterized the endomembrane proteome of switchgrass coleoptiles. They detected 1750 proteins. This dataset is the first proteomic analysis of switchgrass material. To further investigate the contribution of this proteome dataset, they selected for analysis an enzyme essential for cell wall biosynthesis, the UDP-xylose synthase (UXS). Using UXS protein sequences from Arabidopsis and rice, they detected seven UXS proteins in the endo-membrane proteome of switchgrass [73].

Microorganisms in Bioenergy Production

Obtaining energy from any organic substrate is only possible due to microorganisms and derivatives [74]. Since 1857, when Louis Pasteur proved the need for yeast in the sugar-to-alcohol conversion process and showed that various types of microorganisms are required in different types of fermentation, significant advances have been made in this area [75]. By 1882, Robert Koch had confirmed the possibility of cultivating microorganisms on substrates and under artificial conditions, and since then many signs of progress in different fields of science, including biotechnology, have been achieved [76].

Nearly 680 million tons of biomass could be produced in the USA each year by 2030. This quantity is enough to generate more than 10 billion gallons of ethanol or 166 billion kilowatt hours of electricity [77]. Thinking of bioenergy production and of recycling the large amount of residues routinely dispensed from agriculture and industries, it is almost impossible not to consider the utilization of microorganisms for converting the energy confined in those "wastes". Microorganisms can be used for biomass production through photosynthesis, or they can help in the degradation and synthesis of compounds necessary for such purposes [74, 78, 79].

Microorganisms are the primary source for obtaining the enzymes used in the degradation processes of the feedstock [80]. Thus, for example, processes such as fermentation and gasification are managed by complex enzymatic interactions involved in degradation and energy recycling [81, 82]. For example, rumen microorganisms increase the anaerobic bioconversion of lignocellulosic biomass and paper waste [83, 84], and the utilization of extremophilic microorganisms improve the bioconversion of crop residues, overcoming limitations of lignocellulose bioconversions carried out at < 50°C [85].

Among the enzymes commonly used for obtaining energy from the breakdown of the raw material are cellulase, alpha-amylase, protease, glucoamylase, and lipase [81]. Lipases are one of the enzyme types used in biodiesel production, and they act in the degradation of glycerides and fatty acids [86]. For cellulose enzymatic bioconversion, three components are specifically required: endoglucanase, exoglucanase, and β-glucosidase [81]. For lignin degradation, enzymes such as laccase, manganese peroxidase, and lignin peroxidase are provided from Ascomycetes, Deuteromycetes and Basidiomycetes [87, 88].

Different microorganisms, including bacteria, yeasts and fungi, such as *Phanerochaete chrysosporium*, *Trametes versicolor*, *Penicillium restrictum*, *Aspergillus niger*, *Trichoderma* sp., *Candida rugosa*, *Clostridium thermocellum*, *Erwinia chrysanthemi*, *Bacillus* sp., *Pseudomonas fluorescens*, among others, can be used to obtain specific enzymes used for lignocellulosic components or lipid bioconversion [80, 89 - 91]. Therefore, studies of bioenergy production have aimed at biodegradation improvement by identifying new microorganisms and/or using gene insertion to obtain enhanced microorganisms.

The development of biotechnological techniques such as "omics" is essential to provide basic knowledge about the microbial communities involved in bioenergy processes [92]. However, these tools, including proteomics, are dependent on the information available from accurate gene sequences in databases. This is a routine problem regarding the shotgun proteomics of non-model species. It is known, for example, that despite the limitations of techniques before 2008, seventy-five genomes of bioenergy production-related microorganisms had been sequenced and eighty were in process at that time [92]. Currently, thirty-four photosynthetic microorganisms, such as cyanobacteria, have complete genome sequences, demonstrating the increasing interest in this kind of information for advancing proteomics studies [93]. The use of proteomics tools to study microorganisms with potential applications in bioenergy processes involves the identification of populations [94], biologic characterization, genotyping, and the study of relationships with specific substrates, considering the biochemical pathways associated with substrate degradation [79]. The studies reported on below suggest that understanding microorganism-substrate interaction and searching for new enzymes are the priorities in this field.

The genre *Aspergillus* consists of approximately 200 species, some of them known plant pathogens. However, many are important for obtaining organic acids and enzymes used in substrate degradation in industry [80, 94]. The availability of genome sequences of various species of *Aspergillus* sp. has enabled advances in proteomics studies *via* 2-DE, identifying intracellular proteins, cell wall proteins, and sub-proteomes, as well as enabling the characterization of new enzymes such as the heterologues of β-glucosidase. The latter are associated with stress responses and have potential application in industry [80, 95, 96].

Proteome population analyses of *Trichoderma* spp. have identified species that help in the synthesis of enzymes like cellulases [97] and have detected changes in protein synthesis due to the substrate pH in which the fungus grows [98]. Analyses of protein profile during fermentation using *Penicillium chrysogenum* have suggested a correlation among protein profile, biomass concentration, and enzyme production [99].

2D reference maps for comparative studies were produced from organisms such as *Phanerochaete chrysosporium* [100]. Likewise, from different sources of lignocellulose biomass, protein profiles were evaluated initially using the 2D technique with limited protein identification [101, 102], and subsequently using LC-MS/MS, enabling the identification of protein groups related to substrate degradation, which suggested a dependent relationship between protein expression, lignocellulose degradation, and types of agricultural waste [103]. On the other hand, comparative proteomic studies of native and modified strains of *Sacchararomyces cerevisae* for enhanced biofuel precursor production served to identify protein groups that could be related to increased metabolic activity of the modified strain [104].

CONFLICT OF INTEREST

The authors confirm that this chapter content has no conflict of interest.

ACKNOWLEDGEMENTS

None declared.

REFERENCES

[1]   Ndimba BK, Ndimba RJ, Johnson TS, *et al*. Biofuels as a sustainable energy source: an update of the applications of proteomics in bioenergy crops and algae. J Proteomics 2013; 93: 234-44.
[http://dx.doi.org/10.1016/j.jprot.2013.05.041] [PMID: 23792822]

[2]   IRENA, International Renewable Energy Agency. Global Bioenergy: Supply and demand projections A working paper for REmap 2030 2014.

[3]   Witzany G. A perspective on natural genetic engineering and natural genome editing. Introduction. Ann N Y Acad Sci 2009; 1178: 1-5.
[http://dx.doi.org/10.1111/j.1749-6632.2009.05021.x] [PMID: 19845624]

[4]   Wilkins M. Proteomics data mining. Expert review of proteomics 2009; 6(6): 599-603.
[http://dx.doi.org/10.1586/epr.09.81]

[5]   Bukhari SF, Arshad S, Azooz MM, Kazi AG. Omics approaches and abiotic stress tolerance in legumes. In: Azooz MM, Ahmad P, Eds. Legumes under Environmental Stress: Yield, Improvement and Adaptations. Chichester, UK: John Wiley & Sons Ltd 2015.
[http://dx.doi.org/10.1002/9781118917091.ch13]

[6]   Larance M, Lamond AI. Multidimensional proteomics for cell biology. Nat Rev Mol Cell Biol 2015; 16(5): 269-80.
[http://dx.doi.org/10.1038/nrm3970] [PMID: 25857810]

[7]   Agrawal GK, Pedreschi R, Barkla BJ, *et al*. Translational plant proteomics: a perspective. J Proteomics 2012; 75(15): 4588-601.
[http://dx.doi.org/10.1016/j.jprot.2012.03.055] [PMID: 22516432]

[8]   Boggess MV, Lippolis JD, Hurkman WJ, *et al*. The need for agriculture phenotyping: moving from genotype to phenotype. J Proteomics 2013; 93: 20-39.
[http://dx.doi.org/10.1016/j.jprot.2013.03.021] [PMID: 23563084]

[9]   Shoemaker CE, Bransby DI. The Role of Sorghum as a Bioenergy Feedstock. In: Braun R, Karlen D, Johnson D, Eds. Sustainable Alternative Fuel Feedstock Opportunities, Challenges and Roadmaps for Six US Regions. Soil and Water Conservation Society 2008.

[10]   Calviño M, Messing J. Sweet *sorghum* as a model system for bioenergy crops. Curr Opin Biotechnol 2012; 23(3): 323-9.
[http://dx.doi.org/10.1016/j.copbio.2011.12.002] [PMID: 22204822]

[11]   Ngara R, Rees J, Ndimba BK. Establishment of sorghum cell suspension culture system for proteomics studies. Afr J Biotechnol 2008; 7: 744-9.

[12]   Ngara R, Ndimba BK. Mapping and characterisation of the sorghum cell suspension culture secretome. Afr J Biotechnol 2011; 10: 253-66.

[13]   Ngara R, Ndimba R, Borch-Jensen J, Jensen ON, Ndimba B. Identification and profiling of salinity stress-responsive proteins in Sorghum bicolor seedlings. J Proteomics 2012; 75(13): 4139-50.
[http://dx.doi.org/10.1016/j.jprot.2012.05.038] [PMID: 22652490]

[14]   Swami AK, Alam SI, Sengupta N, Sarin R. Differential proteomic analysis of salt stress response in Sorghum bicolor leaves. Environ Exp Bot 2011; 71: 321-8.
[http://dx.doi.org/10.1016/j.envexpbot.2010.12.017]

[15]   Sekhwal MK, Swami AK, Sarin R, Sharma V. Identification of salt treated proteins in sorghum using gene ontology linkage. Physiol Mol Biol Plants 2012; 18(3): 209-16.
[http://dx.doi.org/10.1007/s12298-012-0121-y] [PMID: 23814435]

[16]   Jedmowski C, Ashoub A, Beckhaus T, Berberich T, Karas M, Brüggemann W. Comparative Analysis of Sorghum bicolor Proteome in Response to Drought Stress and following Recovery. Int J Proteomics 2014; 395906.

[17]   Saykhedkar S, Ray A, Ayoubi-Canaan P, Hartson SD, Prade R, Mort AJ. A time course analysis of the extracellular proteome of *Aspergillus nidulans* growing on sorghum stover. Biotechnol Biofuels 2012; 5(1): 52.
[http://dx.doi.org/10.1186/1754-6834-5-52] [PMID: 22835028]

[18]   Che.avegatti-Gianotto. Sugarcane (*Saccharum X officinarum*): A Reference Study for the Regulation of Genetically Modified Cultivars. Brazil Tropical Plant Biol 2011; 4: 62-89.

[19]   Alexander AG. The Energy Cane Alternative. Amsterdam, The Netherlands: Elsevier 1985.

[20]   Matsuoka S, Kennedy AJ, Dos Santos EG, Tomazela AL, Rubio LC. Energy Cane: Its Concept, Development, Characteristics, and Prospects Advances in Botany. 2014.

[21]   Butterfield MK, D'Hont A, Berding N. The sugarcane genome: a synthesis of current understanding, and lessons for breeding and biotechnology. In: Proceedings of the South African Sugar Technologists' Association (SASTA '01); Vol. 75 Durban, South Africa. 2001; pp. : 1-5.

[22]   Menossi M, Silva-Filho MC, Vincentz M, Van-Sluys M-A, Souza GM. Sugarcane Functional Genomics: Gene Discovery for Agronomic Trait Development. Int J Plant Genomics 2008; 2008: 458732.

[23]   Amalraj RS, Selvaraj N, Veluswamy GK, *et al.* Sugarcane proteomics: establishment of a protein extraction method for 2-DE in stalk tissues and initiation of sugarcane proteome reference map. Electrophoresis 2010; 31(12): 1959-74.
[http://dx.doi.org/10.1002/elps.200900779] [PMID: 20564692]

[24]   Zhou G, Yang LT, Li YR, *et al.* Proteomic analysis of osmotic stress-response proteins in sugarcane leaves. Plant Mol Biol Rep 2012; 30: 349-59.
[http://dx.doi.org/10.1007/s11105-011-0343-0]

[25]   Pacheco CM, Pestana-Calsa MC, Gozzo FC, Mansur Custodio Nogueira RJ, Menossi M, Calsa T Jr. Differentially delayed root proteome responses to salt stress in sugar cane varieties. J Proteome Res 2013; 12(12): 5681-95.
[http://dx.doi.org/10.1021/pr400654a] [PMID: 24251627]

[26]   Cesarino I, Araújo P, Sampaio Mayer JL, Paes Leme AF, Mazzafera P. Enzymatic activity and proteomic profile of class III peroxidases during sugarcane stem development. Plant Physiol Biochem 2012; 55: 66-76.
[http://dx.doi.org/10.1016/j.plaphy.2012.03.014] [PMID: 22551762]

[27]   Cesarino I, Araújo P, Paes Leme AF, Creste S, Mazzafera P. Suspension cell culture as a tool for the characterization of class III peroxidases in sugarcane. Plant Physiol Biochem 2013; 62: 1-10.
[http://dx.doi.org/10.1016/j.plaphy.2012.10.015] [PMID: 23159486]

[28]   Brown RL, Menkir A, Chen Z-Y, *et al.* Breeding aflatoxin-resistant maize lines using recent advances in technologies - a review. Food Addit Contam Part A Chem Anal Control Expo Risk Assess 2013; 30(8): 1382-91.
[http://dx.doi.org/10.1080/19440049.2013.812808] [PMID: 23859902]

[29]   Solomon BD, Birchler J, Goldman SL, Zhang Q. Basic information on maize Compendium of bioenergy plants: corn. Boca Raton: CRC Press 2014; pp. 1-32.

[30]   Jayasundara S, Wagner-Riddle C, Dias G, Kariyapperuma KA. Energy and Greenhouse Gas Intensity of Corn (*Zea Mays* L.) Production in Ontario: A Regional Assessment. Can J Soil Sci 2014; 94(1): 77-95.
[http://dx.doi.org/10.4141/cjss2013-044]

[31]   Ye Z, Zheng Y, Li B, Borrusch MS, Storms R, Walton JD. Enhancement of synthetic Trichoderma-based enzyme mixtures for biomass conversion with an alternative family 5 glycosyl hydrolase from *Sporotrichum thermophile*. PLoS One 2014; 9(10): e109885.
[http://dx.doi.org/10.1371/journal.pone.0109885] [PMID: 25295862]

[32]   Li M, Heckwolf M, Crowe JD, *et al.* Cell-wall properties contributing to improved deconstruction by alkaline pre-treatment and enzymatic hydrolysis in diverse maize (*Zea mays L.*) lines. J Exp Bot 2015; 66(14): 4305-15.
[http://dx.doi.org/10.1093/jxb/erv016] [PMID: 25871649]

[33]   Pechanova O, Takáč T, Samaj J, Pechan T. Maize proteomics: an insight into the biology of an important cereal crop. Proteomics 2013; 13(3-4): 637-62.
[http://dx.doi.org/10.1002/pmic.201200275] [PMID: 23197376]

[34]   Gao D, Chundawat SP, Krishnan C, Balan V, Dale BE. Mixture optimization of six core glycosyl hydrolases for maximizing saccharification of ammonia fiber expansion (AFEX) pretreated corn stover. Bioresour Technol 2010; 101(8): 2770-81.
[http://dx.doi.org/10.1016/j.biortech.2009.10.056] [PMID: 19948399]

[35]   Silva-Sanchez C, Chen S, Zhu N, Li Q-B, Chourey PS. Proteomic comparison of basal endosperm in maize miniature1 mutant and its wild-type Mn1. Front Plant Sci 2013; 4: 211.
[PMID: 23805148]

[36]   Robbins ML, Roy A, Wang P-H, *et al.* Comparative proteomics analysis by DIGE and iTRAQ provides insight into the regulation of phenylpropanoids in maize. J Proteomics 2013; 93: 254-75.
[http://dx.doi.org/10.1016/j.jprot.2013.06.018] [PMID: 23811284]

[37]   Brockerhoff EG, Jactel H, Parrotta JA, Ferraz SFB. Role of Eucalypt and Other Planted Forests in Biodiversity Conservation and the Provision of Biodiversity-Related Ecosystem Services. For Ecol Manage 2013; 301: 34-50.
[http://dx.doi.org/10.1016/j.foreco.2012.09.018]

[38]   Nieminen K, Robischon M, Immanen J, Helariutta Y. Towards optimizing wood development in bioenergy trees. New Phytol 2012; 194(1): 46-53.
[http://dx.doi.org/10.1111/j.1469-8137.2011.04011.x] [PMID: 22474686]

[39]   Hinchee M, Rottmann W, Mullinax L, *et al.* Short-Rotation Woody Crops for Bioenergy and Biofuels Applications. Vitr Cell Dev Biol - Plant 2009; 45(6): 619-29.
[http://dx.doi.org/10.1007/s11627-009-9235-5]

[40]   Parsons HT, Weinberg CS, Macdonald LJ, *et al.* Golgi enrichment and proteomic analysis of developing Pinus radiata xylem by free-flow electrophoresis. PLoS One 2013; 8(12): e84669.
[http://dx.doi.org/10.1371/journal.pone.0084669] [PMID: 24416096]

[41]   Kalluri UC, Hurst GB, Lankford PK, Ranjan P, Pelletier DA. Shotgun proteome profile of Populus developing xylem. Proteomics 2009; 9(21): 4871-80.
[http://dx.doi.org/10.1002/pmic.200800854] [PMID: 19743414]

[42]   Loziuk PL, Parker J, Li W, *et al.* Elucidation of Xylem-Specific Transcription Factors and Absolute Quantification of Enzymes Regulating Cellulose Biosynthesis in *Populus trichocarpa*. J Proteome Res 2015; 14(10): 4158-68.
[http://dx.doi.org/10.1021/acs.jproteome.5b00233] [PMID: 26325666]

[43]   Chhetri AB, Tango MS, Budge SM, Watts KC, Islam MR. Non-edible plant oils as new sources for biodiesel production. Int J Mol Sci 2008; 9(2): 169-80.
[http://dx.doi.org/10.3390/ijms9020169] [PMID: 19325741]

[44]   Liu H, Yang Z, Yang M, Shen S. The differential proteome of endosperm and embryo from mature seed of *Jatropha curcas*. Plant Sci 2011; 181(6): 660-6.
[http://dx.doi.org/10.1016/j.plantsci.2011.03.012] [PMID: 21958708]

[45]   Loei H, Lim J, Tan M, *et al.* Proteomic analysis of the oil palm fruit mesocarp reveals elevated oxidative phosphorylation activity is critical for increased storage oil production. J Proteome Res 2013; 12(11): 5096-109.
[http://dx.doi.org/10.1021/pr400606h] [PMID: 24083564]

[46]   Teh HF, Neoh BK, Wong YC, *et al.* Hormones, polyamines, and cell wall metabolism during oil palm fruit mesocarp development and ripening. J Agric Food Chem 2014; 62(32): 8143-52.
[http://dx.doi.org/10.1021/jf500975h] [PMID: 25032485]

[47]   Gupta D, Eldakak M, Rohila JS, Basu C. Biochemical analysis of kerosene tree *Hymenaea courbaril* L. under heat stress. Plant Signal Behav 2014; 9(10): e972851.
[http://dx.doi.org/10.4161/15592316.2014.972851] [PMID: 25482765]

[48]   Wasiak S, Legendre-Guillemin V, Puertollano R, *et al.* Enthoprotin: a novel clathrin-associated protein identified through subcellular proteomics. J Cell Biol 2002; 158(5): 855-62.
[http://dx.doi.org/10.1083/jcb.200205078] [PMID: 12213833]

[49]   Kamal AH, Cho K, Choi JS, *et al.* The wheat chloroplastic proteome. J Proteomics 2013; 93: 326-42.
[http://dx.doi.org/10.1016/j.jprot.2013.03.009] [PMID: 23563086]

[50]   Dreger M, Bengtsson L, Schöneberg T, Otto H, Hucho F. Nuclear envelope proteomics: novel integral membrane proteins of the inner nuclear membrane. Proc Natl Acad Sci USA 2001; 98(21): 11943-8.
[http://dx.doi.org/10.1073/pnas.211201898] [PMID: 11593002]

[51]   Parsons HT, Christiansen K, Knierim B, *et al.* Isolation and proteomic characterization of the Arabidopsis Golgi defines functional and novel components involved in plant cell wall biosynthesis. Plant Physiol 2012; 159(1): 12-26.
[http://dx.doi.org/10.1104/pp.111.193151] [PMID: 22430844]

[52]   Garin J, Diez R, Kieffer S, *et al.* The phagosome proteome: insight into phagosome functions. J Cell Biol 2001; 152(1): 165-80.
[http://dx.doi.org/10.1083/jcb.152.1.165] [PMID: 11149929]

[53]   Salvato F, Havelund JF, Chen M, *et al.* The potato tuber mitochondrial proteome. Plant Physiol 2014; 164(2): 637-53.
[http://dx.doi.org/10.1104/pp.113.229054] [PMID: 24351685]

[54]   Bayer EM, Bottrill AR, Walshaw J, *et al.* Arabidopsis cell wall proteome defined using multidimensional protein identification technology. Proteomics 2006; 6(1): 301-11.
[http://dx.doi.org/10.1002/pmic.200500046] [PMID: 16287169]

[55]   Millar AH, Taylor NL. Subcellular proteomics-where cell biology meets protein chemistry. Front Plant Sci 2014; 5: 55.
[http://dx.doi.org/10.3389/fpls.2014.00055] [PMID: 24616726]

[56]   Kleffmann T, Russenberger D, von Zychlinski A, *et al.* The *Arabidopsis thaliana* chloroplast proteome reveals pathway abundance and novel protein functions. Curr Biol 2004; 14(5): 354-62.
[http://dx.doi.org/10.1016/j.cub.2004.02.039] [PMID: 15028209]

[57]   Heazlewood JL, Tonti-Filippini JS, Gout AM, Day DA, Whelan J, Millar AH. Experimental analysis of the Arabidopsis mitochondrial proteome highlights signaling and regulatory components, provides assessment of targeting prediction programs, and indicates plant-specific mitochondrial proteins. Plant Cell 2004; 16(1): 241-56.
[http://dx.doi.org/10.1105/tpc.016055] [PMID: 14671022]

[58]   Sweetlove LJ, Taylor NL, Leaver CJ. Isolation of intact, functional mitochondria from the model plant *Arabidopsis thaliana.* Methods Mol Biol 2007; 372: 125-36.
[http://dx.doi.org/10.1007/978-1-59745-365-3_9] [PMID: 18314722]

[59]   Eubel H, Lee CP, Kuo J, Meyer EH, Taylor NL, Millar AH. Free-flow electrophoresis for purification of plant mitochondria by surface charge. Plant J 2007; 52(3): 583-94.
[http://dx.doi.org/10.1111/j.1365-313X.2007.03253.x] [PMID: 17727614]

[60]   Ferro M, Salvi D, Brugière S, *et al.* Proteomics of the chloroplast envelope membranes from *Arabidopsis thaliana.* Mol Cell Proteomics 2003; 2(5): 325-45.
[PMID: 12766230]

[61]   Zhao Q, Chen S, Dai S. C4 photosynthetic machinery: insights from maize chloroplast proteomics. Front Plant Sci 2013; 4(85): 85.
[PMID: 23596450]

[62]   Grimaud F, Renaut J, Dumont E, *et al.* Exploring chloroplastic changes related to chilling and freezing tolerance during cold acclimation of pea (*Pisum sativum* L.). J Prot 2013; 80: 145-59.

[63]   Majeran W, Cai Y, Sun Q, van Wijk KJ. Functional differentiation of bundle sheath and mesophyll maize chloroplasts determined by comparative proteomics. Plant Cell 2005; 17(11): 3111-40.
[http://dx.doi.org/10.1105/tpc.105.035519] [PMID: 16243905]

[64]   Majeran W, Friso G, Ponnala L, *et al.* Structural and metabolic transitions of C4 leaf development and differentiation defined by microscopy and quantitative proteomics in maize. Plant Cell 2010; 22(11): 3509-42.
[http://dx.doi.org/10.1105/tpc.110.079764] [PMID: 21081695]

[65]   Covshoff S, Majeran W, Liu P, Kolkman JM, van Wijk KJ, Brutnell TP. Deregulation of maize C4 photosynthetic development in a mesophyll cell-defective mutant. Plant Physiol 2008; 146(4): 1469-81.
[http://dx.doi.org/10.1104/pp.107.113423] [PMID: 18258693]

[66]   Majeran W, van Wijk KJ. Cell-type-specific differentiation of chloroplasts in C4 plants. Trends Plant Sci 2009; 14(2): 100-9.
[http://dx.doi.org/10.1016/j.tplants.2008.11.006] [PMID: 19162526]

[67]   Asif MH, Mantri SS, Sharma A, *et al.* Complete sequence and organisation of the *Jatropha curcas* (Euphorbiaceae) chloroplast genome. Tree Genet Genomes 2010; 6(6): 941-52.
[http://dx.doi.org/10.1007/s11295-010-0303-0]

[68]   Pinheiro CB, Shah M, Soares EL, *et al.* Proteome analysis of plastids from developing seeds of *Jatropha curcas* L. J Prot Res 2013; 12(11): 5137-45.

[69]   Komatsu S, Yanagawa Y. Cell wall proteomics of crops. Front Plant Sci 2013; 4: 17.
[http://dx.doi.org/10.3389/fpls.2013.00017] [PMID: 23403621]

[70]   Wei H, Brunecky R, Donohoe BS, *et al.* Identifying the ionically bound cell wall and intracellular glycoside hydrolases in late growth stage Arabidopsis stems: implications for the genetic engineering of bioenergy crops. Front Plant Sci 2015; 6: 315.
[http://dx.doi.org/10.3389/fpls.2015.00315] [PMID: 26029221]

[71]   Blanch HW, Adams PD, Andrews-Cramer KM, Frommer WB, Simmons BA, Keasling JD. Addressing the need for alternative transportation fuels: the Joint BioEnergy Institute. ACS Chem Biol 2008; 3(1): 17-20.
[http://dx.doi.org/10.1021/cb700267s] [PMID: 18205287]

[72]   Ito J, Petzold CJ, Mukhopadhyay A, Heazlewood JL. The role of proteomics in the development of cellulosic biofuels. Curr Prot 2010; 7: 121-34.
[http://dx.doi.org/10.2174/157016410791330543]

[73]   Lao J, Sharma MK, Sharma R, *et al.* Proteome profile of the endomembrane of developing coleoptiles from switchgrass (*Panicum virgatum*). Proteomics 2015; 15(13): 2286-90.
[http://dx.doi.org/10.1002/pmic.201400487] [PMID: 25677556]

[74]   Hallenbeck PC. Bioenergy from Microorganisms: An Overview, in Microbial BioEnergy: Hydrogen Production In: Zannoni D, De Philippis R, Eds. Advances in Photosynthesis and Respiration; Dordrecht. Netherlands: Springer 2014; 38: pp. 3-21.

[75]   Gal J, Cintas P. Early history of the recognition of molecular biochirality. Top Curr Chem 2013; 333: 1-40.
[http://dx.doi.org/10.1007/128_2012_406] [PMID: 23274573]

[76]   Kaufmann SH, Schaible UE. 100th anniversary of Robert Kochs Nobel Prize for the discovery of the tubercle bacillus. Trends Microbiol 2005; 13(10): 469-75.
[http://dx.doi.org/10.1016/j.tim.2005.08.003] [PMID: 16112578]

[77]   Union of Concerned Scientists (UCS). The promise of biomass: Clean power and fuel—if handled right, Cambridge, MA 2012. Available at: http://www.ucsusa.org/assets/documents/clean_vehicles/BiomassResource-Assessment.pdf

[78]   Tsygankov A, Kosourov S. Immobilization of Photosynthetic Microorganisms for Efficient Hydrogen Production. 2014.
[http://dx.doi.org/10.1007/978-94-017-8554-9_14]

[79]   Kalluri UC, Keller M. Bioenergy research: a new paradigm in multidisciplinary research. J R Soc Interface 2010; 7(51): 1391-401.
[http://dx.doi.org/10.1098/rsif.2009.0564] [PMID: 20542958]

[80]   Polizeli ML, Corrêa EC, Polizeli AM, Jorge JA. Hydrolases from Microorganisms Used for Degradation of Plant Cell Wall and Bioenergy. In: Buckeridge MS, Goldman GH, Eds. Routes to Cellulosic Ethanol. New York, NY: Springer New York 2011; pp. 115-34.
[http://dx.doi.org/10.1007/978-0-387-92740-4_8]

[81]   Serpa VI, Polikarpov I. Enzymes in Bioenergy. In: Buckeridge MS, Goldman GH, Eds. Routes to Cellulosic Ethanol. New York, NY: Springer New York 2011; pp. 97-113.
[http://dx.doi.org/10.1007/978-0-387-92740-4_7]

[82]   Moreira LR. Milanezi NvG, Filho EXF. Enzymology of Plant Cell Wall Breakdown: An Update. In: Buckeridge MS, Goldman GH, Eds. Routes to Cellulosic Ethanol. New York, NY: Springer New York 2011; pp. 73-96.
[http://dx.doi.org/10.1007/978-0-387-92740-4_6]

[83]   Yue Z-B, Li W-W, Yu H-Q. Application of rumen microorganisms for anaerobic bioconversion of lignocellulosic biomass. Bioresour Technol 2013; 128: 738-44.
[http://dx.doi.org/10.1016/j.biortech.2012.11.073] [PMID: 23265823]

[84]   Baba Y, Tada C, Fukuda Y, Nakai Y. Improvement of methane production from waste paper by pretreatment with rumen fluid. Bioresour Technol 2013; 128: 94-9.
[http://dx.doi.org/10.1016/j.biortech.2012.09.077] [PMID: 23196227]

[85]   Bhalla A, Bansal N, Kumar S, Bischoff KM, Sani RK. Improved lignocellulose conversion to biofuels with thermophilic bacteria and thermostable enzymes. Bioresour Technol 2013; 128: 751-9.
[http://dx.doi.org/10.1016/j.biortech.2012.10.145] [PMID: 23246299]

[86]   Ribeiro BD, de Castro AM, Coelho MAZ, Freire DMG. Production and Use of Lipases in Bioenergy: A Review from the Feedstocks to Biodiesel Production. Enzyme Res 2011; 2011: 615803.
[http://dx.doi.org/10.4061/2011/615803]

[87]   Mayer AM, Staples RC. Laccase: new functions for an old enzyme. Phytochemistry 2002; 60(6): 551-65.
[http://dx.doi.org/10.1016/S0031-9422(02)00171-1] [PMID: 12126701]

[88]   Shraddha Shekher R, Sehgal S, Kamthania M, Kumar A. Laccase: Microbial Sources, Production, Purification, and Potential Biotechnological Applications. Enzyme Res 2011.

[89]   Bak JS. Lignocellulose Depolymerization Occurs *via* an Environmentally Adapted Metabolic Cascades in the Wood-Rotting Basidiomycete Phanerochaete Chrysosporium. Microbiol open 2015; 4(1): 151-66.

[90]   Píva GA, Thomas RW. Biomass and Laccase Production by *Trametes Versicolor*, *Trametes Villosa* and *Pycnoporus Sanguineus*. Int Biodeterior Biodegradation 1996; 37(1-2): 119.
[http://dx.doi.org/10.1016/0964-8305(96)84326-5]

[91]　Zeng J, Singh D, Gao D, Chen S. Effects of lignin modification on wheat straw cell wall deconstruction by Phanerochaete chrysosporium. Biotechnol Biofuels 2014; 7(1): 161.
[http://dx.doi.org/10.1186/s13068-014-0161-3] [PMID: 25516769]

[92]　Rittmann BE, Krajmalnik-Brown R, Halden RU. Pre-genomic, genomic and post-genomic study of microbial communities involved in bioenergy. Nat Rev Microbiol 2008; 6(8): 604-12.
[http://dx.doi.org/10.1038/nrmicro1939] [PMID: 18604223]

[93]　Cyanobase. Available at: http://genome.microbedb.jp/cyanobase/#list-of-species.

[94]　Welker M. Proteomics for routine identification of microorganisms. Proteomics 2011; 11(15): 3143-53.
[http://dx.doi.org/10.1002/pmic.201100049] [PMID: 21726051]

[95]　Culleton H, McKie V, de Vries RP. Physiological and molecular aspects of degradation of plant polysaccharides by fungi: what have we learned from Aspergillus? Biotechnol J 2013; 8(8): 884-94.
[http://dx.doi.org/10.1002/biot.201200382] [PMID: 23674519]

[96]　Kniemeyer O. Proteomics of eukaryotic microorganisms: The medically and biotechnologically important fungal genus Aspergillus. Proteomics. 2011; 11(15): 3232-43.
[http://dx.doi.org/10.1002/pmic.201100087] [PMID: 21726053]

[97]　Pandey S, Srivastava M, Shahid M, *et al.* Trichoderma Species Cellulases Produced by Solid State Fermentation. J Data Mining Genomics Proteomics 2015; 6(2): 1-4.

[98]　Adav SS, Ravindran A, Chao LT, Tan L, Singh S, Sze SK. Proteomic analysis of pH and strains dependent protein secretion of *Trichoderma reesei*. J Proteome Res 2011; 10(10): 4579-96.
[http://dx.doi.org/10.1021/pr200416t] [PMID: 21879708]

[99]　Helmel M, Posch A, Herwig C, Allmaier G, Marchetti-Deschmann M. Proteome Profiling Illustrated by a Large-Scale Fed-Batch Fermentation of *Penicillium Chrysogenum*. EuPA Open Proteom. 2014; 4: 113-20.
[http://dx.doi.org/10.1016/j.euprot.2014.06.002]

[100]　Ozcan S, Yildirim V, Kaya L, *et al. Phanerochaete chrysosporium* soluble proteome as a prelude for the analysis of heavy metal stress response. Proteomics 2007; 7(8): 1249-60.
[http://dx.doi.org/10.1002/pmic.200600526] [PMID: 17366474]

[101]　Sato S, Liu F, Koc H, Tien M. Expression Analysis of Extracellular Proteins from *Phanerochaete Chrysosporium* Grown on Different Liquid and Solid Substrates. Microbiol 2007; 153(Pt 9): 3023-33.
[http://dx.doi.org/10.1099/mic.0.2006/000513-0]

[102]　Ravalason H, Jan G, Mollé D, *et al.* Secretome analysis of *Phanerochaete chrysosporium* strain CIRM-BRFM41 grown on softwood. Appl Microbiol Biotechnol 2008; 80(4): 719-33.
[http://dx.doi.org/10.1007/s00253-008-1596-x] [PMID: 18654772]

[103]　Adav SS, Ravindran A, Sze SK. Quantitative proteomic analysis of lignocellulolytic enzymes by *Phanerochaete chrysosporium* on different lignocellulosic biomass. J Proteomics 2012; 75(5): 1493-504.
[http://dx.doi.org/10.1016/j.jprot.2011.11.020] [PMID: 22146477]

[104]　Tang X, Feng H, Zhang J, Chen WN. Comparative proteomics analysis of engineered *Saccharomyces cerevisiae* with enhanced biofuel precursor production. PLoS One 2013; 8(12): e84661.
[http://dx.doi.org/10.1371/journal.pone.0084661] [PMID: 24376832]

CHAPTER 7

Genomics as a Tool for Bioenergy and Biofuel Crops

Sabrina D. Soares, Mariane B. Sobreiro, Vanessa C. Araújo and **Evandro Novaes**[*]

Escola de Agronomia, Universidade Federal de Goiás, Goiânia, Brazil

Abstract: The quest for a renewable and inexpensive source of energy is one of the greatest challenges of the 21st century. Plants have already been used as a renewable source of energy, and continue to be one of the greatest hopes in this area. In order for plants to continue providing a cost-effective and renewable source of energy, it is imperative that their biomass growth and quality be constantly improved. Recent advances in the genomics field, led by the development of high-throughput sequencing and genotyping platforms, have opened up new strategies for accelerating plant breeding and aiding biotechnology development. Analyses of the vast amount of data generated by these modern genomics platforms are only possible with the constant development of computers and bioinformatics tools. In this chapter, we will present the genomic resources available for the most important plant species with bioenergy potential. Bioinformatics tools for gene expression analyses with RNA-Seq and for SNP genotyping are also presented.

Keywords: Bioinformatics, Gene expression, Plants, Phytozome, RNA-seq, Sugarcane, SNP genotyping.

INTRODUCTION

One of the greatest challenges of the 21st century revolves around finding a renewable and inexpensive source of energy [1]. Energy is essential for world economic development and for sustaining our modern and ever growing conveniences (transportation, air conditioning, appliances, *etc.*). A renewable energy source is urgently needed, given the disastrous effects of climate change we and other species have been facing. There is a substantial body of scientific evidence linking climate change to the relatively recent and extensive emissions of CO_2 from the use of non-renewable fossil fuels [2].

[*] **Corresponding author Evandro Novaes:** Federal University of Goiás, Goiás-GO, Brazil; Tel: +55-62-3521-1687; Fax: +55-62-3521-1600; E-mail: novaes_ufg@yahoo.com

It is estimated that by the year 2035, carbon dioxide emissions will be increased from 31 gigatons (Gt) to 37 Gt [3]. Distributed, cost-effective, and renewable sources of energy are thus resources that are much needed in order to face the challenges posed by climate change. Currently, renewable energies account for only 13% of total energy consumption worldwide, with bioenergy representing 10% of this value. Bioenergy derives from biomass (biological raw material), in the form of solid, liquid, or gas products [4].

Plants have already been used as a renewable source of energy, and they continue to be one of the greatest hopes in relation to this matter. These organisms have the capacity to absorb CO_2 from the atmosphere and to store sunlight energy through photosynthesis. The energy accumulated in their biomass can be released by burning or can be converted into much needed liquid biofuel [5–7].

In order for plants to continue providing a cost-effective, renewable source of energy, it is imperative that their biomass growth and quality be constantly improved. Traditional plant breeding has proven to be a successful method of doing so [8]. However, it is considered to be a slow process, especially with perennial species such as forest trees, for example, the fast-growing species of the *Eucalyptus* and *Pinus* genera.

With recent advances in the genomics field, led by the development of high-throughput sequencing and genotyping platforms, scientists have been focused on a strategy for accelerating the breeding process. Genomic selection is a method that uses thousands of molecular markers, well distributed throughout the genome, to predict the breeding value of genotypes from a breeding population [9]. By having a well-adjusted genomic selection model, breeders can predict the value of a genotype very early, as soon as the plant produces its first leaves for DNA extraction (in the nursery). With this technology, breeders can thus perform early selection and faster advance the breeding cycle. The power of genomic selection has its greatest potential, arguably, in perennial plant species [10].

Sequencing and functional analyses of plant genomes are also very powerful for discovering genes and/or regulatory elements for biotechnology. This is especially important for the challenge of converting biomass into biofuels, where plant cell wall degradation is still a major challenge that can potentially be met with biotechnology [2].

In this chapter, we will present: a) the genomic resources available for the most important species with bioenergy potential; b) Phytozome as a genomics database of bioenergy crops; and c) some bioinformatics tools for exploiting the genomic resources available for bioenergy crops. More specifically, we will show bioinformatics pipelines for gene expression analyses with RNA-Seq and for SNP genotyping with the Genome Analyses Toolkit [11].

PLANT SPECIES WITH BIOENERGY AND BIOFUEL POTENTIAL

The main crops used globally for bioethanol production include corn (produced primarily in the United States), sugarcane (Brazil and South Africa), beets (European Union), and wheat (European Union and China). For biodiesel production, the main species are canola and sunflower (European Union), soy (United States and Brazil), and palm oil (Southeast Asia) [12, 13].

A comparative study on the productivity of corn and sugarcane as raw materials for bioenergy at different latitudes has shown that sugarcane produces on average three times more energy per hectare than corn [6]. The planted area in Brazil has increased from 1.4 to 7 million hectares between 1960 and 2007 [14]. Even with this high-efficiency and large sugarcane planted area, Brazil still lags behind the US, the largest producer of biofuel. Production in the US is largely based on corn. After Brazil, the European Union is the third largest producer, mostly based on biodiesel production from rapeseed and sunflower [4].

In addition to starch and sugar for the production of first generation biofuels, sugarcane bagasse, wood, corncobs, and other plant residues are also potential sources (lignocellulosic biomass) for energy production. Energy derived from this type of biomass, that is, from the cell wall instead of starch or sugar, is classified as a second generation technology [6]. This biomass originates from non-food lignocellulosic materials such as woody energy crops (eucalyptus, poplar, alfalfa, reed canary grass, elephant grass, switchgrass, among others), agricultural wastes (wheat husk, stems, cobs), forest residues (cuttings, wood fuel), and wood processing residues, as well as urban and industrial solid wastes (such as paper and cardboard) [15 - 18]. New policies stipulate that bioenergy must be generated from different biomass combinations, preferentially including non-food crops, to avoid the potential of increasing food prices. Next generation fuel production will enable middle and long-term solutions for sustaining economic development in a scenario of global climate changes [18].

GENOMIC RESOURCES AVAILABLE FOR PLANT SPECIES WITH BIOENERGY AND BIOFUEL POTENTIAL

There is a significant effort on the part of agriculture to develop more efficient crops to meet global demand. The search for more efficient crops to produce bioenergy and biofuels requires advances in agronomics and plant breeding. The daunting challenge is to select plants, with potential as energy crops, that possess increased productive efficiency. Among the characteristics needed to enhance plants with bioenergy potential are increased biomass growth, response to light competition, branching, stem thickness, the chemical composition of the cell wall, as well as tolerance to biotic and abiotic stresses, especially water deficit. The aim is to maximize biomass productivity and minimize the global impact on land use [19].

The desired cell wall composition of bioenergy crops depends on the final use of the biomass. If energy is obtained simply by burning the biomass, generally a high lignin content in the cell walls is desired. Lignin is a highly energetic molecule [20]. On the other hand, lignin is undesired if the biomass is to be used for second generation biofuel production, especially ethanol from cellulose. Lignin is covalently bound to the cellulose molecules, making extraction and fermentation of cellulose monomers (glucose) a difficult task [21].

Table **1** shows the main energy crops with genome sequences available or being generated. These species can provide raw material from sources such as sugar, starch, and oleaginous seeds. Some also exhibit great potential as raw materials for second generation energy, *via* lignocellulosic biomass.

Table 1. List of plants showing bioenergy potential, with their genome sequenced or with sequencing underway.

Raw material	Species	Popular name	Genome	Author cited	Data bank
Starch and sugar	*Beta vulgaris*	Beet	Yes	Dohm *et al.*, 2014 [30]	NCBI
	Manihot esculenta	Cassava	Yes	Prochnik *et al.*, 2012 [31]	Phytozome
	Zea mays	Corn	Yes	Schnable *et al.*, 2009 [24]	Phytozome
	Triticum aestivum	Wheat	Yes	Marcussen *et al.*, 2014 [32]	Phytozome
	Sorghum bicolor	Sorghum	Yes	Paterson *et al.*, 2009 [25]	Phytozome
	Brassica rapa	Canola	*	-	Phytozome http://brassicadb.org
Oleaginous seed	*Glycine max*	Soy	Yes	Schmutz *et al.*, 2010 [26]	Phytozome
	Elaeis guineensis	Palm oil	Yes	Singh *et al.*, 2013 [33]	NCBI
	Helianthus (genus)	Sunflower	*	-	http://sunflowergenome.org

(Table 1) contd.....

Raw material	Species	Popular name	Genome	Author cited	Data bank
Lignocellulose	*Eucalyptus grandis*	Eucalyptus	Yes	Myburg *et al.*, 2014 [28]	Phytozome
	Populus trichocarpa	Poplar	Yes	Tuskan *et al.*, 2006 [29]	Phytozome
	Panicum virgatum	Elephant grass	*	-	Phytozome

* Species that are still in the sequencing data analysis phase. Information is available at the links indicated.

Table **1** also demonstrates that the vast majority of species with bioenergy potential have already undergone genome sequencing. Two notable exceptions are sugarcane (*Saccharum* spp.) and pine (*Pinus* spp.). Even though these species are very important for bioenergy production, their large and complex genomes are impairing their full characterization in a timely manner. However, given the economic interest, these remaining species are in the process of having their genomes sequenced and assembled [22 - 29]. In addition to genome sequencing, all of these species have molecular marker technology and genetic maps available, as well as an extensive number of genetic studies on natural and breeding populations. Also available is the analysis of gene expression in these bioenergy crops under different environmental conditions and ontogenetic stages.

The use of genomic resources is important in the genetic enhancement of these species [19]. Genome sequencing facilitates the development and genotyping of molecular markers. The availability of thousands of markers, covering the entire genome, is essential to aiding technologies such as genome-wide selection for accelerated breeding. Additionally, genome sequence availability can facilitate molecular genetics and molecular biology studies that identify genes with biotechnology potential, such as those used for increasing oil production in *Brassica napus*.

SUGARCANE: A BIOENERGY SPECIES WITH A LARGE, POLYPLOID, COMPLEX GENOME

Sugarcane (*Saccharum* spp.) is one of the main crops in biofuel production (ethanol). It is an indigenous (domesticated) plant and can be grown in tropical and subtropical regions on both sides of the equator [34]. Bioethanol produced from sugarcane has a positive energy balance, since sugarcane growth absorbs more carbon than is emitted when its ethanol is burned as fuel [35]. Sugarcane is a C4 plant that is highly efficient in converting sunlight into biomass [36]. Its production cost is low when compared to other crops, as it can re-sprout (ratoon) after harvest and therefore does not require replanting every growing season [14, 37].

Moreover, it is very efficient as a source of saccharose for the production of first generation biofuels. As stated earlier, intense efforts are being made to use the residues obtained after saccharose extraction. This has prompted an increase in research on second generation ethanol, since sugarcane bagasse exhibits high lignocellulosic content that can be used to generate bioenergy [38]. The bioenergy contained in the bagasse is already being used to produce the heat and energy required to convert sugar into bioethanol [39]. In Brazil, the excess energy produced by the mills is converted into electricity and sold to feed the electrical grid.

The genome of sugarcane is large, polyploid, highly complex, and contains a large proportion (35-45%) of repetitive elements [40 - 42]. With these characteristics, sugarcane has arguably the most complex genome of cultivated plant species. This complexity has hindered any attempt to perform a comprehensive characterization of the sugarcane genome. Many advances have been made, however, such as the sequencing of euchromatic gene-enriched portions of the genome [42].

Given the highly complex nature of the sugarcane genome, it is not surprising that a lot of efforts have concentrated on studying transcriptome and gene expression differences under different genotypes, environmental conditions, and stresses [43 - 46]. Genes that affect these characteristics can be used as targets for biotechnology [19].

PHYTOZOME: A GENOME SEQUENCE DATABASE FOR BIOENERGY CROPS

The Phytozome website (http://phytozome.jgi.doe.gov/pz/portal.html) is a repository of plant genome sequences [47]. Currently, it contains the genome sequences of more than 50 plant species, a number of which exhibit bioenergy potential (Table **1**). Phytozome is a genomic database of plants constructed and maintained by the US Department of Energy's Joint Genome Institute (DOE/JGI). The DOE/JGI has sequenced most of the plant genomes available to date. This high level of investment is in line with the mission of the JGI "to develop renewable and sustainable sources of biofuels from plant biomass, and to understand the biological processes controlling greenhouse gas accumulation in the atmosphere".

In addition to genome sequences, Phytozome contains gene annotations produced by JGI bioinformaticians, as well as Genome Browsers for all species with sequenced genomes. Furthermore, BLAST and BLAT sequence search algorithms have been incorporated into the database, making it possible to search within a species' genome or genes, and also within the genomes of an entire taxonomic group. The Phytozome database also depicts the homology and ancestral relationships between proteins, making it possible to identify homologue gene families in different species. The database is frequently updated as to genomes and gene annotations. The current version is 11.0, with fifty-seven genomes sequenced and annotated.

BIOINFORMATICS TOOLS GREW TOGETHER WITH THE DNA SEQUENCING PLATFORMS

The primary objective of bioinformatics is the computational analysis of molecular sequences such as DNA, RNA, and proteins. These molecules are the main focus of molecular biology. Computers emerged as important tools for studies in this field in the early 1960s [48]. Automatic sequencers were developed in the 1990s, thereby generating a large amount of data for analysis and storage. The first animal genome sequenced was *Caenorhabditis elegans* in 1998, and in the year 2000 the first plant with genome sequencing was *Arabidopsis thaliana* [49]. Powerful computers and rapid computational tools were required to assemble, characterize, and search the sequences obtained. DNA sequencing was thus the primary driver of bioinformatics.

The science of bioinformatics involves different fields of knowledge, including mathematics, statistics, software engineering, computer sciences, and molecular biology. It consists of all types of study or computational tools that can organize and/or generate new biological information. Computational molecular biology encompasses the area of bioinformatics, where computational and information science methodologies are used within genomics, mapping, sequencing, and structure determination. Using these methods, it is possible to obtain, maintain, and visualize large amounts of biological data, answering important biological questions and proposing new hypotheses [50].

The development of high-performance sequencing platforms has been accompanied by the rapid evolution of new bioinformatics tools. Bioinformatics is absolutely necessary for analyzing the large amounts of data generated by these platforms. One of the first sequencing methods was described by Frederick Sanger. In 1977, he and Alan Coulson published two studies that led to a revolution in biology [51, 52]. Their pioneering research introduced a methodology capable of resolving the sequence of an entire genome. They sequenced all the 5386 nucleotides of the single-stranded bacteriophage φX174 [51]. This method underwent a number of changes that culminated in the creation of the first automated DNA sequencer, the ABI 370 (Applied Biosystems), in 1986. In 1995, microarrays and serial analysis of gene expression (SAGE) provided the opportunity for efficiently assessing the expression of thousands of genes [53]. High-throughput analyses of gene expression were important for functional analyses of genes and for improving genome annotation.

The publication of human genome sequences [54, 55], opened up the genomics era. The promise was to apply genomic information in medicine and agricultural practice. With this hope, a big market for high-throughput sequencing platforms was created. As a consequence, next generation sequencers (NGS) such as the 454 and Illumina/Solexa platforms emerged in 2005 [53]. Even though many next generation sequencing platforms have been created since 2005, Illumina has consistently been that with the largest proportion of the market. The sequencing method underlying its platforms was developed by the Solexa Company and was later acquired and improved by Illumina. Illumina's market success is due to the large volume of sequencing data generated by its method, associated with high accuracy and low cost per base pair [56]. The large volume of data produced makes it possible for the Illumina method to be applied in a wide array of studies, such as genomic and transcriptomic characterization, gene expression using RNA-Seq, polymorphism genotyping, methylome and epigenetics studies, and analyses of protein-DNA interaction *via* Chip-Seq.

Next generation sequencing uses methodologies that overcome the main difficulties of the Sanger method, that is, the need for cloning DNA fragments and for electrophoresis to perform sequencing. These next generation platforms have thus revolutionized the scientific fields that can use DNA or RNA sequences, with advantages that include high yield, sensitivity, and speed [57].

Sequencing technologies include steps such as DNA-template preparation, sequential incorporation of new nucleotides, image capture, data sequencing, and analysis. The combination of specific protocols differentiates one technology from another and determines the type of data produced on each platform. For example, while 454 and Ion Torrent use emulsion PCR to clone DNA fragments bound to beads, the Illumina method involves a technique called bridge PCR to clone the DNA fragments hybridized to probes on the surface of a sequencing slide. With these differences, the biological applications, costs, time, and quality of data obtained also differ among platforms [56]

Another possibility for using the NGS platform is RNA sequencing, thereby obtaining the transcriptome information of an organism. This methodology is known as RNA-seq, enabling the study of gene expression and splicing events [58] and the identification of transcriptome complexity with high accuracy [59].

Sequencing and bioinformatics tools, increasingly effective for genome studies, have provided the opportunity for studying differential gene expression. These studies can identify genes and metabolic pathways, important in creating biotechnologies that improve the yield and quality of biomass used for biofuels and bioenergy. Moreover, these next generation platforms have also made it possible to genotype thousands of single-nucleotide polymorphisms (SNPs). Current plant improvement programs have taken advantage of this easy assessment of genetic variability at the genome level, in order to create predictive models of the genotypic value of plants and their breeding populations. This technology, known as genomic selection, has the potential to make the production of biofuel and bioenergy more efficient.

Finally, with respect to bioinformatics, it is important to underscore the Linux operational system. The vast majority of bioinformatics computational tools are used in Linux distributions, depending on computers and programs based on this operational system. Bio-Linux was created to facilitate the use of bioinformatics programs, providing complete software package systems and support systems. Bio-Linux was developed by NERC Environmental Bioinformatics Centre in 2002 [60]. It is used in bioinformatics analysis, and its developers describe it as "an ideal system for scientists handling and analyzing biological data".

DIFFERENTIAL EXPRESSION ANALYSES WITH RNA-SEQ

Genomic democratization has occurred with the advent and popularization of next generation sequencing technologies. This is due to lower costs per sequenced base and higher data yield.

Among the possible applications of next generation sequencers is the transcriptome study of a species, that is, the study of the total set of transcripts of a particular cell, tissue, or whole organism at a given moment [61]. The RNA-seq technique is defined as the use of NGS platforms for analyzing the gene expression of all of an organism's genes under different conditions (tissues, treatments, ontogenetic stage, *etc.*). The method involves the transformation of target RNAs into complementary or cDNA molecules, in order to identify and quantify them *via* sequencing in a given tissue, organ, and/or ontogenetic stage.

The RNA-seq technique can be used to study relative transcript abundance and to identify new transcripts such as non-coding RNAs, alternative transcripts (isoforms), and polymorphisms (such as SNPs), in genic regions [62].

The aim of gene expression quantification studies is to observe the expression profile in a given situation, or to compare transcriptional response patterns under different conditions [63]. Once the sequences (reads) are obtained, the first bioinformatics step is to map them onto a genome or transcriptome of reference. If no reference is available, the reads should be assembled and then mapped onto the assembled transcriptome. Trinity [64] is a program commonly used to reconstruct (assemble) transcriptome sequences.

Next, the level of transcript or isoform expression is estimated using a count table containing information on the number of times a transcript was identified in each of the treatments or conditions studied [65]. After the transcripts are counted, the reads are normalized in order to control for different biases inherent in the counting process or created by the sequencing technique [66]. Each program can have its own normalization mechanism. However, the most common normalization methods are: trimmed mean of M-values (TMM) [67]; reads per kilobase of exon model per million mapped reads (RPKM) [68]; and fragments per kilobase of exon per million fragments mapped (FPKM) [69]. Regardless of the method, the idea is to obtain bias-free measures of gene expression that are not influenced by differences in gene length or by the number of sequences obtained for the different treatments.

Analyses of differential expression are conducted after counting and data normalization in order to determine whether the intergroup (treatments) differences are significant. The most widely applied mathematical models for studying gene expression using sequence counting are based on the Poisson and the negative binomial distributions. These distributions are more appropriate for modeling sampling differences between technical and biological replicates, respectively [70].

Fig. (**1**) shows a typical bioinformatics pipeline for RNA-Seq studies. However, the protocol to follow will depend on the experimental design and on the dataset structure, that is: 1) if there are replicates or not; 2) if there are more than two sample groups; and, finally, 3) if the replicates are technical or biological. Furthermore, it is important to determine if the expression levels of the different isoforms are relevant for the situation in question [71]. Not every software is capable of modeling and detecting expression differences between isoforms.

SNP DISCOVERY AND GENOTYPING

Molecular marker genotyping is at the heart of applied genomics. By having thousands of molecular markers genotyped in an animal or plant breeding population, breeders can fit a model to predict the genotypic or breeding value of individuals [72]. These genome-wide selection (GWS) models have the potential to accelerate the breeding cycles of perennial bioenergy crops, as demonstrated for *Eucalyptus* [10, 73] and *Pinus* [73] species. In addition to advancing breeding and research with bioenergy crops, molecular marker genotyping is also having a profound impact in the fields of medicine, ecology, and conservation [74, 75]. As these impacts depend on thousands of markers, single nucleotide polymorphisms (SNPs) have been the marker of choice for high genome coverage. SNPs are the most abundant type of polymorphism in any organism, whether prokaryotes or eukaryotes.

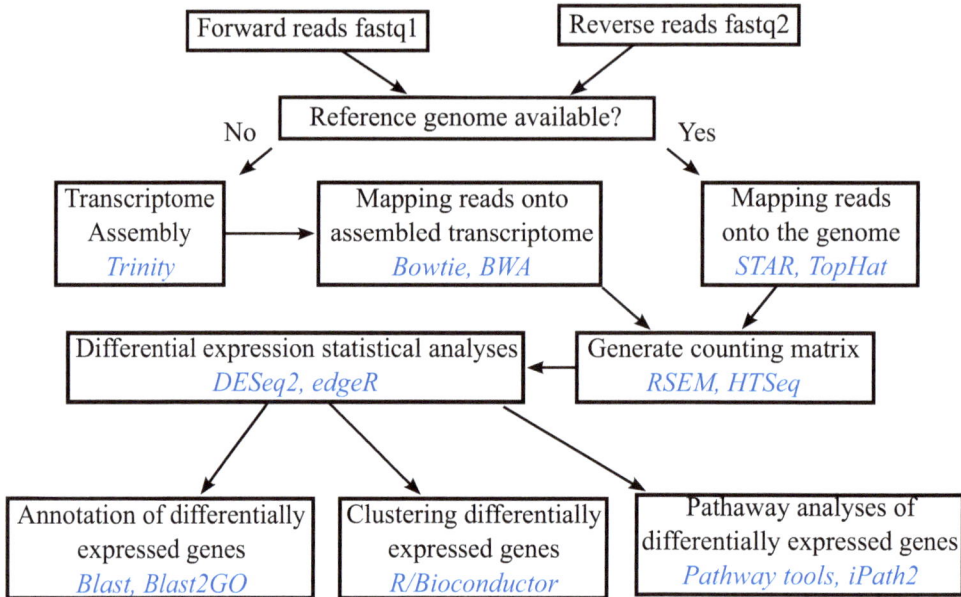

Fig. (1). Bioinformatics pipeline for RNA-Seq analyses of gene expression. For each step, commonly used programs are shown in blue. The pipelines branch out in the beginning. If there is a genome sequence available, sequences can be assembled into the genome with a splice-aware program (*e.g.*, STAR or TopHat). If there is no genome sequence available, RNA sequences can be assembled into transcripts with Trinity. In this case, RNA sequences can be mapped into the assembled transcripts, for example with Bowtie or BWA. After mapping a gene expression, a count matrix can be obtained by counting the number of sequences overlapping each gene or transcript. Based on this matrix, containing the number of sequences mapped in each gene with data from each library (treatment), DESeq2 or edgeR can be used to estimate the significance of the gene expression differences among treatments. The list of differentially expressed genes can be annotated, grouped with cluster analyses, and visualized within biochemical pathways.

SNPs can be genotyped with a myriad of technologies [76]. Technologies can be classified into low- and high-throughput methods. Methods dependent on PCR, such as SNIP-SNP [77] and TaqMan [78], are generally low-throughput, genotyping only up to a dozen SNPs in parallel. Methods based on mass spectrometry are considered of intermediate throughput, allowing parallel genotyping of dozens (and sometimes more than forty) SNPs simultaneously [79].

However, as there is a broad market for genome-wide genotyping applications, many high-throughput SNP genotyping methods have been developed.

Microarrays were the first platform to allow genotyping of many thousands of SNPs in parallel [80]. The method is based on the fact that polymorphisms, including SNPs, can interfere or disrupt hybridization of DNA fragments onto microarray probes. For many years, microarrays have been the method of choice for high-throughput SNP genotyping [81, 82]. However, microarrays have limitations for SNP genotyping. First, they depend on a genome sequence reference for probe design, although DArT technology [83, 84] can circumvent this limitation by using spotted arrays with randomly cloned genomic fragments used as probes. Second, cross-hybridization between homologue sequences, especially between paralogous regions of the genome, can increase the number of false positive (paralogous) SNPs. Third, with complex eukaryotic genomes there is a need for complexity reduction in order to obtain reasonable signal-to-noise ratios on hybridizations. This reduced genome representation can be achieved, for example, with exome capture technologies [85]. Finally, there may be more than one polymorphism within the probe sequence. If these close SNPs are not in linkage disequilibrium, their genotyping is noisy and convoluted.

To circumvent some of these problems, Illumina has combined the high-parallelization property of microarrays with the high specificity and signal-t--noise obtained with fluorescent-based, single-base extension SNP detection. The technology is called Infinium and the detection is performed on a modified microarray called BeadArray [86]. Currently, Illumina's Infinium technology is the gold standard method for high-throughput SNP detection. It is being applied in human disease association studies [87], and in natural and breeding populations of plants, including those with bioenergy potential such as eucalyptus [88], maize [89], and soybean [90]. Affymetrix has also developed improved microarrays (Axiom technology) for SNP detection [91], that have been used in bioenergy plant species [92, 93]. Even though these improved microarray technologies (Axiom and Infinium) have remained the gold standard for SNP genotyping, they have the disadvantage of requiring extensive sequencing data to assemble a reference genome and to identify thousands of SNP markers.

As the cost of DNA sequencing continues to plummet, genotype by sequencing methods [94] are increasing in popularity. The advantage is that SNP discovery and genotyping can be performed simultaneously, without any knowledge about the genomic positions of SNPs. The main disadvantage is the required computational and data storage infrastructure, in addition to the bioinformatics knowledge necessary for performing analyses with the massive number of sequences generated on next generation (NGS) platforms [95]. Another problem of SNP genotyping with sequencing data lies in the relatively high sequencing error of NGS platforms. To reduce uncertainty, it is desirable that SNPs are genotyped only in regions with high sequencing depth [96]. Higher depth can be obtained by sequencing a smaller proportion of the genome, either by using methods based on restriction enzymes and PCR [94], or by using sequencing-capture probes [97].

The bioinformatics pipelines used to identify SNPs from next generation sequencing data usually involve the following steps: 1) mapping reads onto the reference genome sequence; 2) SNP genotyping; 3) quality control and SNP filtration; 4) analyses and annotation of the most significant SNPs. There are many bioinformatics tools available as options for performing each step. For the first mapping step, two commonly used and highly efficient software programs are Bowtie [98] and BWA [99]. SNP genotyping based on next generation sequencing data can be performed with software such as the Genome Analyses Toolkit (GATK) [100], FreeBayes [101], VarScan [102], and Samtools mpileup [103]. GATK is an interesting option, as it performs recalibration of the sequencing quality scores as an additional step for avoiding false positive SNPs [104]. Fig. (**2**) shows a typical pipeline for genotyping SNPs with GATK.

Quality control can be performed within the genotyping software or with additional software. For example, GATK has a number of filtering options with its VariantFiltration script. Another interesting option is to use VCFTools [105] that can filter SNPs based on a number of parameters, such as minimum allele frequency (MAF), sequencing depth of reference and alternative alleles, quality score of the genotyping call, *etc*. VCFTools can also output the transition to the transversion (Ts/Tv) ratio of the identified SNPs. In humans and many plant species, the Ts/Tv ratio is on average higher than 2.0. As sequencing error tends to be random, at least in most cases, it generally decreases the Ts/Tv ratio since there are four possible transversions (A-T, A-C, G-T, G-C) for two transitions (A-G, C-T).

Once SNPs are identified, the genotyping table, which is generally in the variant call format (VCF) [105], can be used to perform a number of analyses depending on how the individuals were sampled. If they were sampled from natural or breeding populations, they could be used for population genomics studies, for association mapping, for genomic selection, *etc*. If the analyses find specific loci, such as those with selection signature from population genomics, or those associated with quantitative traits from association studies, the most significant SNPs can be annotated in terms of where they are in the genome. SnpEff [106] is a software that indicates, based on the genome annotation, whether the SNP falls inside a gene and classifies it in terms of the likelihood of disrupting gene function.

```
┌─────────────────────┐                    ┌─────────────────────┐
│ Forward reads fastq1│                    │ Reverse reads fastq2│
└─────────────────────┘                    └─────────────────────┘
```

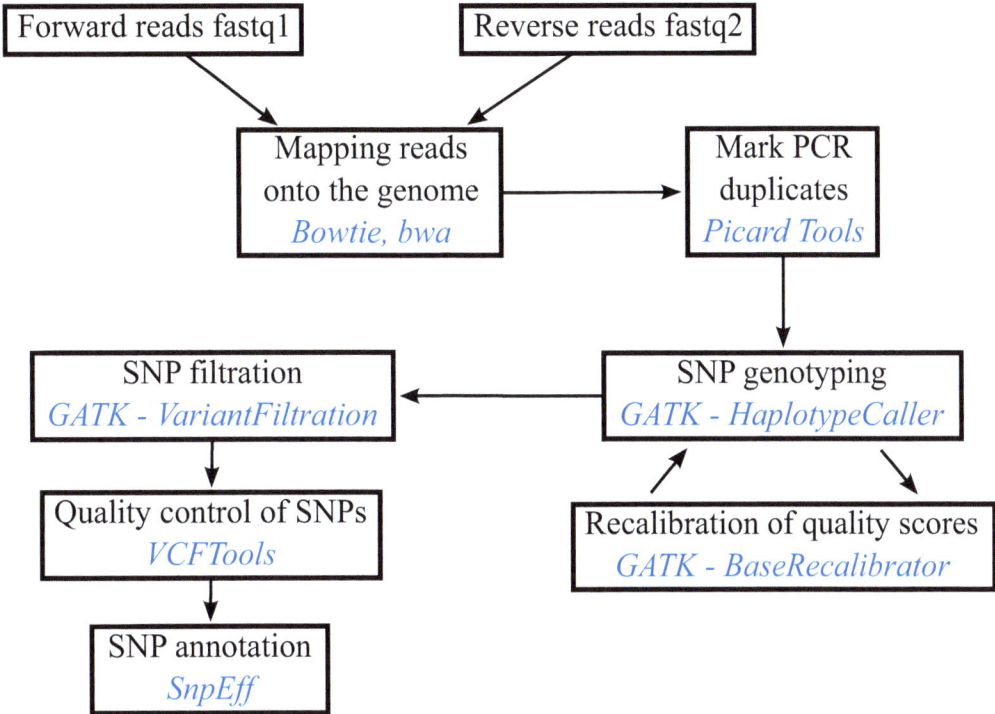

Fig. (2). Bioinformatics pipeline for SNP genotyping, filtration, and annotation. For each step, commonly used programs are shown in blue. Sequencing files (fastq format) are mapped onto the reference genome sequence. Duplicated reads, generally an artifact of PCR amplification, are marked from the alignment file (SAM format). The GATK program is then used to genotype and filter SNPs. A list of "gold standard" SNPs can be used for base recalibration with GATK. For species without previous knowledge of SNPs, the filtered SNPs can be used for quality score recalibration. Finally, SNPs can be filtered with VCFTools and annotated in terms of where they are in relation to the annotated genes with SnpEff.

CONCLUSION

Plant species can surely provide a renewable source of biomass for us to face one of the greatest challenges of the 21st century, that of finding a renewable, inexpensive, and well-distributed source of energy. It is therefore important that plant species with bioenergy potential are constantly improved towards higher biomass productivity and quality for their easy conversion into bioenergy or biofuel. The recent advances in the field of genomics and its platforms, especially in DNA sequencing technologies, have provided new research opportunities for molecular biologists and plant breeders. The vast amount of data generated in the next generation sequencing (NGS) platforms makes it relatively inexpensive to survey the entire transcriptome of a species in any tissue or condition. This has revolutionized the study of gene function, facilitating the development of new biotechnologies. In the same way, these NGS technologies make it easy to genotype thousands to millions of polymorphisms, well distributed throughout the genome, in many natural and breeding populations. This has revolutionized population genetics studies, introducing the field of population genomics, and also has the potential to completely change and accelerate plant breeding. By having thousands of molecular markers throughout the genome, breeders can adjust genomic selection models that can predict the breeding value of individual plants at early developmental stages (in the nursery). This has the potential to substantially accelerate the breeding cycle turn over, especially in perennial species such as forest trees. In addition, fewer experiments have to be conducted in order to identify the best genotypes to be selected. This is an exciting time for geneticists interested in species with bioenergy potential.

CONFLICT OF INTEREST

The authors confirm that this chapter content has no conflict of interest.

ACKNOWLEDGEMENTS

The authors wish to thank the Brazilian funding agencies, Conselho Nacional de Desenvolvimento Científico e Tecnológico (CNPq) and Fundação de Amparo à Pesquisa do Estado de Goiás (FAPEG), for supporting genomics projects 476709/2012-1 and 2012-012-72750053, respectively.

REFERENCES

[1]   Barney JN, Ditomaso JM. Nonnative Species and Bioenergy: Are We Cultivating the Next Invader? Bioscience 2008; 58(1): 64.

[2]   de Siqueira Ferreira S, Nishiyama MY, Paterson AH, *et al.* A Chromosome-Based Draft Sequence of the Hexaploid Bread Wheat (*Triticum Aestivum*) Genome. Nature 2013; 345(6194): 1250092.

[3]   Ipcc. Summary for Policymakers. Clim Chang 2013 Phys. Sci. Basis. Contrib. Work. Gr. I to Fifth Assess. Rep Intergov Panel Clim Chang 2013 2013; 33.

[4]   Eisentraut A. Sustainable production of second-generation biofuels. International Biofuel Agency (IEA) 2010; pp. 45-55. https://www.iea.org/publications/freepublications/publication/second_generation_biofuels.pdf

[5]   Wallace R, Ibsen K, McAloon A, Yee W. Feasibility Study for Co-Locating and Integrating Ethanol Production Plants from Corn Starch and Lignocellulosic Feedstocks. Natl Renew Energy Lab 2005; pp. 1-58.

[6]   Runge CF, Sheehan JJ, Senauer B, *et al.* Assessing the Comparative Productivity Advantage of Bioenergy Feedstocks at Different Latitudes. Environ Res Lett 2012; 7(4): 045906.

[7]   Daianova L, Dotzauer E, Thorin E, Yan J. Evaluation of a Regional Bioenergy System with Local Production of Biofuel for Transportation, Integrated with a CHP Plant. Appl Energy 2012; 92: 739-49.

[8]   Duvick DN. The Contribution of Breeding to Yield Advances in Maize (*Zea Mays L.*); Advances in Agronomy. Elsevier 2005; Vol. 86.

[9]   Wen M, Wang H, Xia Z, Zou M, Lu C, Wang W. Development of EST-SSR and Genomic-SSR Markers to Assess Genetic Diversity in *Jatropha Curcas* L. BMC Res Notes 2010; 3: 42.

[10]   Grattapaglia D, Resende MD. Genomic Selection in Forest Tree Breeding. Tree Genet Genomes 2011; 7(2): 241-55.

[11]   McKenna A, Hanna M, Banks E, *et al.* The Genome Analysis Toolkit: A MapReduce Framework for Analyzing next-Generation DNA Sequencing Data. Genome Res 2010; 20(9): 1297-303.

[12]   Yuan JS, Tiller KH, Al-Ahmad H, Stewart NR, Stewart CN. Plants to Power: Bioenergy to Fuel the Future. Trends Plant Sci 2008; 13(8): 421-9.

[13]   Gupta VK, Potumarthi R, O'Donovan A, Kubicek CP, Sharma GD, Tuohy MG. Bioenergy Research: An Overview on Technological Developments and Bioresources. In: Gupta VK, Kubicek MG, Xu JS, Eds. Bioenergy Research: Advances and Applications. Amsterdam: Elsevier 2014; pp. 23-47.

[14]   Martinelli LA, Filoso S. Expansion of Sugarcane Ethanol Production in Brazil: Environmental and Social Challenges. Ecol Appl 2008; 18(4): 885-98.

[15]   Sanderson KU. Biofuels: A Field in Ferment. Nature 2006; 444(7120): 673-6.

[16]   Jungmeier G, Hingsamer M, Van Ree R. Biofuel-Driven Biorefineries A Selection of the Most Promising Biorefinery Concepts to Produce Huge Volumes of Road Transportation Biofuels until 2025. IEA Bioenergy Task 2013; p. 42.

[17]   Lee RA, Lavoie J-M. From First- to Third-Generation Biofuels: Challenges of Producing a Commodity from a Biomass of Increasing Complexity. Anim Front 2013; 3(2): 6-11.

[18]   Ho DP, Ngo HH, Guo W. A Mini Review on Renewable Sources for Biofuel. Bioresour Technol 2014; 169: 742-9.

[19]   Rubin EM. Genomics of Cellulosic Biofuels. Nature 2008; 454(7206): 841-5.

[20]   Mu W, Ben H, Ragauskas A, Deng Y. Lignin Pyrolysis Components and Upgrading—Technology Review. BioEnergy Res 2013; 6(4): 1183-204.

[21]   Sannigrahi P, Ragauskas AJ, Tuskan GA. Poplar as a Feedstock for Biofuels: A Review of Compositional Characteristics. Biofuels Bioprod Biorefin 2010; 4(2): 209-26.

[22]   Dohm JC, Minoche AE, Holtgräwe D, *et al.* The Genome of the Recently Domesticated Crop Plant Sugar Beet (*Beta Vulgaris*). Nature 2014; 505(7484): 546-9.

[23]   Prochnik S, Marri PR, Desany B, *et al.* The Cassava Genome: Current Progress, Future Directions. Trop Plant Biol 2012; 5(1): 88-94.

[24]   Schnable PS, Ware D, Fulton RS, *et al.* The B73 Maize Genome: Complexity, Diversity, and Dynamics. Science 2009; 326(5956): 1112-5.

[25]   Paterson AH, Bowers JE, Bruggmann R, *et al.* The Sorghum Bicolor Genome and the Diversification of Grasses. Nature 2009; 457(7229): 551-6.

[26]   Schmutz J, Cannon SB, Schlueter J, *et al.* Genome Sequence of the Palaeopolyploid Soybean. Nature 2010; 463(7278): 178-83.

[27]   Singh R, Ong-Abdullah M, Low EL, *et al.* Oil Palm Genome Sequence Reveals Divergence of Interfertile Species in Old and New Worlds. Nature 2013; 500(7462): 335-59.

[28]   Myburg A, Grattapaglia D, Tuskan G, *et al.* The Genome of *Eucalyptus Grandis*. Nature 2014; 509(7505): 356-62.

[29] Tuskan GA, DiFazio S, Jansson S, *et al.* The Genome of Black Cottonwood, *Populus Trichocarpa* (Torr. & Gray). Science 2006; 313(5793): 1596-604.

[30] Dohm JC, Minoche AE, Holtgräwe D, *et al.* The Genome of the Recently Domesticated Crop Plant Sugar Beet (*Beta Vulgaris*). Nature 2014; 505(7484): 546-9.

[31] Prochnik S, Marri PR, Desany B, *et al.* The Cassava Genome: Current Progress, Future Directions. Trop Plant Biol 2012; 5(1): 88-94.

[32] Marcussen T, Sandve SR, Heier L, *et al.* A Chromosome-Based Draft Sequence of the Hexaploid Bread Wheat (*Triticum Aestivum*) Genome. Science 2014; 345(6194): 1250092.

[33] Singh R, Ong-Abdullah M, Low EL, *et al.* Oil Palm Genome Sequence Reveals Divergence of Interfertile Species in Old and New Worlds. Nature 2013; 500(7462): 335-59.

[34] Lisboa CC, Butterbach-Bahl K, Mauder M, Kiese R. Bioethanol Production from Sugarcane and Emissions of Greenhouse Gases - Known and Unknowns. GCB Bioenergy 2011; 3(4): 277-92.

[35] Akalın PK, Han Y, Gao S, *et al.* Chapter 2 - Bioenergy Research: An Overview on Technological Developments and Bioresources. Bioinformatics 2013; 3(4): 1-7.

[36] Casu RE, Dimmock CM, Chapman SC, *et al.* Identification of Differentially Expressed Transcripts from Maturing Stem of Sugarcane by *in Silico* Analysis of Stem Expressed Sequence Tags and Gene Expression Profiling. Plant Mol Biol 2004; 54(4): 503-17.

[37] Del Grosso S, Smith P, Galdos M, Hastings A, Parton W. Sustainable Energy Crop Production. Curr Opin Environ Sustain 2014; 9: 20-5.

[38] Bottcher A, Cesarino I, Santos AB, *et al.* Lignification in Sugarcane: Biochemical Characterization, Gene Discovery, and Expression Analysis in Two Genotypes Contrasting for Lignin Content. Plant Physiol 2013; 163(4): 1539-57.

[39] Carroll A, Somerville CR. Cellulosic Biofuels. Annu Rev Plant Biol 2009; 60: 165-82.

[40] Garsmeur O, Charron C, Bocs S, *et al.* High Homologous Gene Conservation despite Extreme Autopolyploid Redundancy in Sugarcane. New Phytol 2011; 189(2): 629-42.

[41] Figueira TR, Okura V, Rodrigues da Silva F, *et al.* A BAC Library of the SP80-3280 Sugarcane Variety (*saccharum Sp.*) and Its Inferred Microsynteny with the Sorghum Genome. BMC Res Notes 2012; 5(1): 185.

[42] Grativol C, Regulski M, Bertalan M, *et al.* Sugarcane Genome Sequencing by Methylation Filtration Provides Tools for Genomic Research in the Genus Saccharum. Plant J 2014; 79(1): 162-72.

[43] Carnavale Bottino M, Rosario S, Grativol C, *et al.* High-Throughput Sequencing of Small RNA Transcriptome Reveals Salt Stress Regulated microRNAs in Sugarcane. PLoS One 2013; 8(3): e59423.

[44] Cardoso-Silva CB, Costa EA, Mancini MC, *et al. De Novo* Assembly and Transcriptome Analysis of Contrasting Sugarcane Varieties. PLoS One 2014; 9(2): e88462.

[45] Casu RE, Rae AL, Nielsen JM, Perroux JM, Bonnett GD, Manners JM. Tissue-Specific Transcriptome Analysis within the Maturing Sugarcane Stalk Reveals Spatial Regulation in the Expression of Cellulose Synthase and Sucrose Transporter Gene Families. Plant Mol Biol 2015; 89(6): 607-28.

[46] Vicentini R, Bottcher A, Brito MD, *et al.* Large-Scale Transcriptome Analysis of Two Sugarcane Genotypes Contrasting for Lignin Content. PLoS One 2015; 10(8): e0134909.

[47] Goodstein DM, Shu S, Howson R, *et al.* Phytozome: A Comparative Platform for Green Plant Genomics. Nucleic Acids Res 2012; 40(D1): 1178-86.

[48] Hagen JB. The Origins of Bioinformatics. Nat Rev Genet 2000; 1(3): 231-6.

[49] Initiative TA. Analysis of the Genome Sequence of the Flowering Plant *Arabidopsis Thaliana*. Nature 2000; 408(6814): 796-815.

[50]   Akalın PK. Introduction to Bioinformatics. Mol Nutr Food Res 2006; 50(7): 610-9.

[51]   Sanger F, Air GM, Barrell BG, *et al.* Nucleotide Sequence of Bacteriophage Phi X174 DNA. Nature 1977; 265(5596): 687-95.

[52]   Sanger F, Nicklen S, Coulson AR. DNA Sequencing with Chain-Terminating Inhibitors. Proc Natl Acad Sci USA 1977; 74(12): 5463-7.

[53]   Morozova O, Hirst M, Marra MA. Applications of New Sequencing Technologies for *Transcriptome Analysis.* Annu Rev Genomics Hum Genet 2009; 10(1): 135-51.

[54]   Lander ES, Linton LM, Birren B, *et al.* Initial Sequencing and Analysis of the Human Genome. Nature 2001; 409(6822): 860-921.

[55]   Venter C, Adams M, Myers E, *et al.* The Sequence of the Human Genome. Science 2001; 291(5507): 1304-51.

[56]   Metzker ML. Sequencing Technologies - the next Generation. Nat Rev Genet 2010; 11(1): 31-46.

[57]   Han Y, Gao S, Muegge K, Zhang W, Zhou B. Advanced Applications of RNA Sequencing and Challenges. Bioinform Biol Insights 2015; 9: 29-46.

[58]   Varuzza L. Introdução à Análise de Dados de Sequenciadores de Nova Geração. Leonardo Varuzza's Site 2013; pp. 1-76.

[59]   Marguerat S, Bähler J. RNA-Seq: From Technology to Biology. Cell Mol Life Sci 2010; 67(4): 569-79.

[60]   Field D, Tiwari B, Snape J. Bioinformatics and Data Management Support for Environmental Genomics. PLoS Biol 2005; 3(8): 1352-3.

[61]   Wang Z, Gerstein M, Snyder M. RNA-Seq: A Revolutionary Tool for Transcriptomics. Nat Rev Genet 2009; 10(1): 57-63.

[62]   Sims D, Sudbery I, Ilott NE, Heger A, Ponting CP. Sequencing Depth and Coverage: Key Considerations in Genomic Analyses. Nat Rev Genet 2014; 15(2): 121-32.

[63]   Finotello F, Di Camillo B. Measuring Differential Gene Expression with RNA-Seq: Challenges and Strategies for Data Analysis. Brief Funct Genomics 2015; 14(2): 130-42.

[64]   Grabherr MG, Haas BJ, Yassour M, *et al.* Trinity: Reconstructing a Full-Length Transcriptome without a Genome from RNA-Seq Data. Nat Biotechnol 2013; 29(7): 644-52.

[65]   Zhang ZH, Jhaveri DJ, Marshall VM, *et al.* A Comparative Study of Techniques for Differential Expression Analysis on RNA-Seq Data. PLoS One 2014; 9(8): e103207.

[66]   Oshlack A, Robinson MD, Young MD. From RNA-Seq Reads to Differential Expression Results. Genome Biol 2010; 11(12): 220.

[67]   Robinson MD, Oshlack A. A Scaling Normalization Method for Differential Expression Analysis of RNA-Seq Data. Genome Biol 2010; 11(3): R25.

[68]   Mortazavi A, Williams BA, McCue K, Schaeffer L, Wold B. Mapping and Quantifying Mammalian Transcriptomes by RNA-Seq. Nat Methods 2008; 5(7): 621-8.

[69]   Trapnell C, Williams BA, Pertea G, *et al.* Transcript Assembly and Quantification by RNA-Seq Reveals Unannotated Transcripts and Isoform Switching during Cell Differentiation. Nat Biotechnol 2010; 28(5): 511-5.

[70]   Soneson C, Delorenzi M. A Comparison of Methods for Differential Expression Analysis of RNA-Seq Data. BMC Bioinformatics 2013; 14(1): 91.

[71]   Khang T, Lau C. Getting the most out of RNA-seq data analysis. Peer J 32015; : e1360. https://www.ncbi.nlm.nih.gov/pmc/articles/PMC4631466/pdf/peerj-03-1360.pdf

[72]   Meuwissen TH, Hayes BJ, Goddard ME. Prediction of Total Genetic Value Using Genome-Wide Dense Marker Maps. Genetics 2001; 157(4): 1819-29.

[73] Resende MF, Munoz P, Resende MD, *et al.* Accuracy of Genomic Selection Methods in a Standard Data Set of Loblolly Pine (*Pinus Taeda L.*). Genetics 2012; 190(4): 1503-10.

[74] Manolio TA. Bringing Genome-Wide Association Findings into Clinical Use. Nat Rev Genet 2013; 14(8): 549-58.

[75] Narum SR, Buerkle CA, Davey JW, Miller MR, Hohenlohe PA. Genotyping-by-Sequencing in Ecological and Conservation Genomics. Mol Ecol 2013; 22(11): 2841-7.

[76] Syvanen AC. Toward Genome-Wide SNP Genotyping. Nat Genet 2005; 37(6) (Suppl.).

[77] Wicks SR, Yeh RT, Gish WR, Waterston RH, Plasterk RH. Rapid Gene Mapping in *Caenorhabditis Elegans* Using a High Density Polymorphism Map. Nat Genet 2001; 28(june): 160-4.

[78] Schleinitz D, Distefano JK, Kovacs P. Targeted SNP Genotyping Using the Taqman® Assay. Methods Mol Biol 2011; 700: 77-87.

[79] Millis MP. Medium-Throughput SNP Genotyping Using Mass Spectrometry: Multiplex Snp Genotyping Using the Iplex?? Gold Assay. Methods Mol Biol 2011; 700: 61-76.

[80] Winzeler Ea, Richards DR. Direct Allelic Variation Scanning of the Yeast Genome. Science 1998; 281(5380): 1194-7.

[81] Matsuzaki H, Loi H, Dong S, *et al.* Parallel Genotyping of over 10,000 SNPs Using a One-Primer Assay on a High-Density Oligonucleotide Array. Genome Res 2004; 14(3): 414-25.

[82] Matsuzaki H, Dong S, Loi H, *et al.* Genotyping over 100,000 SNPs on a Pair of Oligonucleotide Arrays. Nat Methods 2004; 1(2): 109-11.

[83] Jaccoud D, Peng K, Feinstein D, Kilian A. Diversity Arrays: A Solid State Technology for Sequence Information Independent Genotyping. Nucleic Acids Res 2001; 29(4): E25.

[84] Sansaloni CP, Petroli CD, Carling J, *et al.* A High-Density Diversity Arrays Technology (DArT) Microarray for Genome-Wide Genotyping in Eucalyptus. Plant Methods 2010; 6: 16.

[85] Clark MJ, Chen R, Lam HY, *et al.* Performance Comparison of Exome DNA Sequencing Technologies. Nat Biotechnol 2011; 29(10): 908-14.

[86] Steemers FJ, Chang W, Lee G, Barker DL, Shen R, Gunderson KL. Whole-Genome Genotyping with the Single-Base Extension Assay. Nat Methods 2006; 3(1): 31-3.

[87] Song H, Ramus SJ, Tyrer J, *et al.* A Genome-Wide Association Study Identifies a New Ovarian Cancer Susceptibility Locus on 9p22.2. Nat Genet 2009; 41(9): 996-1000.

[88] Silva-Junior OB, Faria DA, Grattapaglia D. A Flexible Multi-Species Genome-Wide 60K SNP Chip Developed from Pooled Resequencing of 240 Eucalyptus Tree Genomes across 12 Species. New Phytol 2015; 206(4): 1527-40.

[89] Hufford MB, Xu X, van Heerwaarden J, *et al.* Comparative Population Genomics of Maize Domestication and Improvement. Nat Genet 2012; 44(7): 808-11.

[90] Song Q, Hyten DL, Jia G, *et al.* Development and Evaluation of SoySNP50K, a High-Density Genotyping Array for Soybean. PLoS One 2013; 8(1)

[91] Hoffmann TJ, Kvale MN, Hesselson SE, *et al.* Next Generation Genome-Wide Association Tool: Design and Coverage of a High-Throughput European-Optimized SNP Array. Genomics 2011; 98(2): 79-89.

[92] Unterseer S, Bauer E, Haberer G, *et al.* A Powerful Tool for Genome Analysis in Maize: Development and Evaluation of the High Density 600 K SNP Genotyping Array. BMC Genomics 2014; 15(1): 823.

[93] Lee YG, Jeong N, Kim JH, *et al.* Development, Validation and Genetic Analysis of a Large Soybean SNP Genotyping Array. Plant J 2015; 81(4): 625-36.

[94]   Davey JW, Hohenlohe PA, Etter PD, Boone JQ, Catchen JM, Blaxter ML. Genome-Wide Genetic Marker Discovery and Genotyping Using next-Generation Sequencing. Nat Rev Genet 2011; 12(7): 499-510.

[95]   Kumar S, Banks TW, Cloutier S. SNP Discovery through next-Generation Sequencing and Its Applications. Int J Plant Genomics 2012.

[96]   Nielsen R, Paul JS, Albrechtsen A, Song YS. Genotype and SNP Calling from next-Generation Sequencing Data. Nat Rev Genet 2011; 12(6): 443-51.

[97]   Neves LG, Davis JM, Barbazuk WB, Kirst M. Whole-Exome Targeted Sequencing of the Uncharacterized Pine Genome. Plant J 2013; 75(1): 146-56.

[98]   Langmead B, Trapnell C, Pop M, Salzberg SL. Ultrafast and Memory-Efficient Alignment of Short DNA Sequences to the Human Genome. Genome Biol 2009; 10: R25.

[99]   Li H, Durbin R. Fast and Accurate Long-Read Alignment with Burrows-Wheeler Transform. Bioinformatics 2010; 26(5): 589-95.

[100] DePristo MA, Banks E, Poplin R, *et al.* A Framework for Variation Discovery and Genotyping Using next-Generation DNA Sequencing Data. Nat Genet 2011; 43(5): 491-8.

[101] Garrison E, Marth G. Haplotype-Based Variant Detection from Short-Read Sequencing. arXiv Prepr arXiv12073907 2012; 9.

[102] Koboldt DC, Chen K, Wylie T, *et al.* VarScan: Variant Detection in Massively Parallel Sequencing of Individual and Pooled Samples. Bioinformatics 2009; 25(17): 2283-5.

[103] Li H, Handsaker B, Wysoker A, *et al.* The Sequence Alignment/Map Format and SAMtools. Bioinformatics 2009; 25(16): 2078-9.

[104] Cornish A, Guda C. A Comparison of Variant Calling Pipelines Using Genome in a Bottle as a Reference. BioMed Res Int 2015.

[105] Danecek P, Auton A, Abecasis G, *et al.* The Variant Call Format and VCFtools. Bioinformatics 2011; 27(15): 2156-8.

[106] Cingolani P, Platts A, Wang LL, *et al.* A Program for Annotating and Predicting the Effects of Single Nucleotide Polymorphisms, SnpEff: SNPs in the Genome of D. Melanogaster Strain W 1118; Iso-2; Iso-3. Fly (Austin) 2012; 6(2): 80-92.

CHAPTER 8

Challenges in Biomass Production for Biofuels

Maria J. Calderan-Rodrigues[1,*], Fabio V. Scarpare[2], João L. N. Carvalho[1], Camila Caldana[1], Marina C. M. Martins[1] and Lucia Mattiello[3]

[1] *Laboratório Nacional de Ciência e Tecnologia do Bioetanol (CTBE), Centro Nacional de Pesquisa em Energia e Materiais (CNPEM), Campinas, Brazil*

[2] *Centro de Energia Nuclear na Agricultura (CENA), Universidade de São Paulo (USP), Piracicaba, SP, Brazil, Faculdade de Engenharia Mecânica (FEM), Universidade Estadual de Campinas (UNICAMP), Campinas, SP, Brazil*

[3] *Laboratório de Genoma Funcional, Departamento de Genética, Evolução e Bioagentes, Universidade Estadual de Campinas (UNICAMP), Campinas, Brazil*

Abstract: Plant biomass is the feedstock for biofuel production. Efforts to maximize yield per unit of production area are of crucial importance in meeting the rising demand for renewable energy sources. Plantation, irrigation, use of fertilizers and pesticides, together with harvest, represent the major costs involved in biomass production. In this chapter we give a broad overview of (i) the factors influencing biomass yield (such as water and nutrients) and advances in cultivation technologies, discussing sustainability issues and the link of these practices with industry needs, and (ii) the relation of these conditions to the physiology of energy crops, presenting innovative technologies that can support management decisions. We will focus on sugarcane as a model for bioenergy crops.

Keywords: Agricultural productivity, Bioenergy crops, Biomass accumulation, Energy cane, First and second generation ethanol, Harvest, Innovative technologies, Management, Metabolomics, Photosynthetic rates, Plant breeding, Plant physiology, Soil mechanization, Straw maintenance, Sucrose, Sugarcane bagasse, Sustainability, Tillage, Water and nutrient uptake, Yield.

INTRODUCTION

Agricultural productivity has a strong correlation with local edaphoclimatic conditions and management. Soil quality, *i.e.*, nutrient and water availability, are key indicators for inferring environmental sustainability. They somehow influence

* **Corresponding author Maria J. Calderan-Rodrigues:** Laboratório Nacional de Ciência e Tecnologia do Bioetanol (CTBE), Centro Nacional de Pesquisa em Energia e Materiais (CNPEM), Campinas, Brazil; Tel: 0055 (19) 35175032; E-mail: juliana.calderan@bioetanol.org.br

Daniela Defavari do Nascimento, & William A. Pickering (Eds.)

accumulation. Agricultural management tools can be used to improve soil fertility, and are also capable of increasing the yield harvested. Complementarily, the understanding of how nutrient and water uptake occur until their incorporation into the plant biomass can assist both in management decisions in the field and in biotechnologies development. The discoveries produced should then go beyond the laboratory scale and be applied not only at the industry level, but also in production fields.

Brazilian ethanol production is favored by several conditions, such as vast areas for planting, geographic position, sunlight incidence, and plenteous rainfall. In addition, the economic and environmental concerns that arose in the last century promoted an increase in the use of Brazilian ethanol in the energy matrix worldwide, resulting in environmental, economic, and social gains. The Brazilian National Alcohol Program (Proálcool) contributed to the increase of the sugarcane harvest area. The area increased from 1.7 Mha in the 1970s and to 9.8 Mha in 2014 [1], and the number of ethanol mills has also increased during more than forty years of technology development.

One of the most productive plants worldwide, sugarcane is also considered the most efficient material for first generation ethanol production [2]. In addition, two-thirds of sugarcane's energy potential lies in bagasse and straw [3], materials that can be used to produce so-called second generation ethanol (E2G). However, as is widely known, the costs of sugarcane E2G production are not yet competitive with first generation ethanol. Along with more established strategies such as increasing the efficiency of enzymes, studies from the biomass perspective related to plant fiber content can offer an extra advantage to E2G feasibility. For all of the reasons mentioned above, therefore, this discussion will be focused on sugarcane production in Brazil.

Due to more recent concern about energy sources, and the resulting increased demand for ethanol, sugarcane production is expanding worldwide [2]. However, inherent to the monoculture model, the occupation of large areas associated with certain agricultural practices raises concerns of possible environmental impacts. Different issues should be integrated in order to find a reasonable solution that maximizes field production with minimal negative impacts. Through the tools used in molecular biology, plant genetics, and physiology, it is possible to reach into new technologies and approaches in order to support biomass field production. One of these innovative technologies is metabolomics, an approach that may be able to unravel the link between plant metabolism and biomass accumulation.

In this chapter, the following topics will be addressed: management and other environmental aspects related to plant yield, such as water and nutrients; management decisions used to eliminate possible bottlenecks; how these measures adopted in the field can affect plant physiology and biomass accumulation, and the other way round; how knowledge and innovative technologies in plant physiology can influence field production.

FACTORS INFLUENCING BIOMASS YIELD AND ADVANCES IN CULTIVATION TECHNOLOGIES

Brazilian sugarcane yields increased dramatically from 46 ton.ha^{-1} in 1970 to 83 ton.ha^{-1} in 2010 [4]. The increment can be attributed to large-scale genetic breeding programs supported by the government and the private sector, particularly in the 1970s and 1980s, but also by better agricultural techniques which in some cases are intrinsically related to mechanization.

In São Paulo State, the largest producer with around 370 million tons annually, a semi-mechanized system is used. Mechanical harvesting is used in 83% of the production area, but mechanized planting is used in only around 45% of the area [5]. Despite the advantages of mechanical harvesting, it is claimed that due to harvest losses, soil compaction, and ratoon damage, mechanical harvesting reduces plant productive life and sugarcane yield as compared with the manual harvesting of burned cane [6].

Heavy soil mechanization due to tillage (plowing, disking, and subsoiling) and excessive traffic associated with other agricultural practices such as mulching, irrigation, and fertilization, particularly those in which byproducts or residues are used, have great potential to affect local water resources.

For centuries, sugarcane was harvested manually and the fields were burned to remove sugarcane straw (the tops and dry leaves), driving away snakes and other potentially poisonous animals and facilitating harvesting. In the last decade, the harvest of sugarcane by human cane cutters has been gradually reduced, and was changed to harvesting with the use of machines that maintained the straw on the ground [7]. The green management of sugarcane requires the maintenance of vegetal residues on the ground, resulting in 10-20 tons of dry matter/hectare. At present, research experiments are addressing the advantages of maintaining plant residues on the soil [8]. An example of these studies is illustrated in Fig. (**1**).

The maintenance of sugarcane residues on the ground can enhance nutrient cycling [9], provide higher water-holding capacity [10], higher aggregate stability [11], increased soil organic carbon stocks [12], and can reduce greenhouse gas emissions [13] (Fig. **2**).

Fig. (1). Field trial with four different amounts of straw on soil surface. a) zero; b) 5 ton/ha; c) 10 ton/ha; d) 15 ton/ha.

Fig. (2). Sugarcane residues on the soil surface after the use of harvesting raw sugarcane (left); water maintenance under straw blanket (right).

The agronomic benefits and drawbacks of straw on the ground surface, using mechanical harvesting, may be directly related to the edaphoclimatic characteristics of the production region. Therefore, numerous options for handling this material exist. The minimal amount on the soil to make the sugarcane production chain more sustainable may vary greatly.

In hotter regions, the advantages of using the plant litter coverage to reduce water losses is restricted if the plant material is not thick enough. The effect of mulch on soil conservation is more predominant during the first 90 days, since the sugarcane canopy heavily occupies the environment in order to capture solar radiation succeeding shoot appearance. In addition, the sugarcane harvest period in center-south Brazil takes place from April/May to October/November, *i.e.*, during the dry season, when straw maintenance in early cane harvest may be essential. Furthermore, soil moisture maintenance due to mulching's effect for early cane varieties is important for preventing yield decay, mainly in expansion areas where extended dry spell periods occur frequently. The effect of straw maintenance on sugarcane yield increase, due to soil moisture, has been reported in several studies [14 - 16].

Aside from its impact on yield, straw maintenance is an effective way of controlling soil erosion by reducing negative water impacts when raining [17, 18]. It is worth highlighting that soil erosion is not so pronounced in comparison to food crops such as maize and soybeans, as sugarcane plants spread very rapidly, closing the surface and covering the soil. This limits the negative impacts in the replanting stage, which occurs only every five or six years [8]. Nonetheless, sugarcane soil waste depends on several circumstances, such as slope, rainfall events, and especially mechanization, which can markedly alter soil hydraulic properties due to soil compaction and disturbance from tillage plowing, disking, and subsoiling.

The effects of soil erosion, however, go beyond the loss of fertile land. Eroded sediment can cause siltation and pollution of water bodies downstream from erosion sources. Fiorio and coworkers [19] studied complications related to sugarcane sedimentation in São Paulo State. The transition from burned sugarcane to mechanized harvesting reduced soil erosion and rainfall runoff [20], due to the large volumes of plant residues on the ground surface.

In terms of soil carbon accumulation, Cerri and coworkers [12] surveyed the most relevant research performed in south central Brazil. They state that around 1.5 ton C/hectare/year was the total for the plant residue maintenance in the sugarcane fields in the largest production region in Brazil.

The unburned sugarcane harvesting process provides a more sustainable sugarcane production. Nonetheless, the use of precise soil management is also an important practice for increasing sustainability. Research has shown the benefits of reduced tillage in lowering soil CO_2 emissions [21] and sugarcane production costs [22]. Corroborating these assumptions, Segnini and coworkers [23] isolated the impacts of maintaining tillage operations and keeping the plant litter on ground surface during sugarcane renovation. According to these authors, the adoption of green cane and no-tillage retained more than double the carbon of conventional tillage. Soil carbon is an important player in sugarcane fields, especially in tropical regions, which present weathered soils and high rates of decomposition of organic matter in soil. It is responsible for improving nutrient recycling, soil water retention, and cation exchange capacity. Although sugarcane has a high drought tolerance, water is essential for sugarcane production since it is directly related to yield. Traditionally, Brazilian sugarcane cultivation does not require irrigation. However, water inputs by irrigation or fertigation increase water use efficiency, especially when using high-efficiency irrigation systems [24].

In terms of irrigation, Brazil has 4.46 million irrigated hectares in use, out of a total potential of 29.3 million hectares [25]. Sugarcane irrigation also positively affects yield and increase crop life age, aside from reducing the need for more planted areas and indirectly decreasing transportation distances and costs [26]. On the other hand, possible negative impacts are soil salinization and water resource depletions.

In the Paraná River basin, where most Brazilian sugarcane is grown, water withdrawal increased by around 50% between 2006 and 2010 (from 492.7 to 736.1 m^3s^{-1}). In this context, the water use for irrigation had the highest increase, from 108.1 to 311.4 m^3s^{-1}. In other sugarcane producing countries such as India and Pakistan, the extraction of groundwater exceeded the regeneration progression of aquifers in nature [27]. In Brazil, the most recent agricultural census survey [28] indicates that subsurface water for irrigation supplies 22% of the total equipped area, as compared with around 43% in the rest of the world [29].

Irrigation in the sugarcane sector can be divided into: (i) salvage irrigation, which is done subsequent to planting or harvesting to ensure plant survival; (ii) deficit irrigation, applied in the most critical development phases; and (iii) full irrigation, aiming to replace all evapo-transpiration demand. In salvage irrigation, vinasse is sprayed through machine guns, and therefore it is considered a fertigation practice.

Vinasse is the main effluent of the ethanol industry chain, and it contains high levels of organic load and thus has great potential to cause environmental impacts. When discharged, it can alter water quality. Biological oxygen demand discharges are given in terms of the corresponding quantity of domiciliary sewage produced by an average citizen [30]. Thus the pollution of a modest mill in Brazil can be compared to that of around one and a half million citizens [31].

During 1980s, Brazil applied a law to banish vinasse release into surface waters. The solution was to recycle this effluent back into the sugarcane fields as organic fertilizer, providing a convenient method of wastewater disposal and preventing potential health and environmental hazards. Moreover, this practice represented an economization of resources, as this kind of fertilizer is imported into the country. However, its frequent use may cause several problems.

In São Paulo State, there is a restriction on vinasse application, regulated by the Standard Technique P4.231 [32], which considers crop nutrition needs.

In addition to vinasse, research [33 - 36] presents sewage effluent as a wastewater source for fertigation practice in sugarcane fields, especially when irrigated areas are near urban centers.

Like vinasse, treated domestic sewage use in irrigation has possibly negative impacts, which comprise the permeation of chemical elements into the groundwater and the agglomeration of unwanted materials in the soil [37]. Besides altering the chemical composition of the soil, the use of domestic sewage may cause a change in soil pH, making it more acidic [38]. On the other hand, as reported in several studies, this wastewater fertigation could achieve yield increase [35, 36].

From a general perspective, the use of residues from the bioethanol production chain such as straw, vinasse, and filter cake, among others, offers an excellent opportunity for improving the sustainability of the sugarcane production system. Regarding green cane management, its implementation has the potential to improve the energy performance of sugarcane. However, there is not yet a well-established tradeoff with regard to whether the straw should be used for energy production or left on the soil surface. There is thus an open question about if and how much straw can be removed from cultivated fields without threatening long-term soil quality and crop productivity. In the case of vinasse use, with the technological advances in E2G it is expected that other types of vinasse will be produced with a significantly different composition from that of first generation ethanol (E1G). E2G has higher organic matter content, but its nutrient and mineral content, especially potassium, is considerably lower than that of E1G vinasse, a factor which may further hinder fertigation with E2G vinasse.

PHYSIOLOGY OF ENERGY CROPS AND BIOMASS ACCUMULATION

The interaction of nutrients and soil conditions can influence research related to plant physiology and nutrition. Plant biomass production is closely linked to growth as well as development, and depends on a tight regulation of a complex signaling network integrating molecular, physiological, and biochemical processes with environmental factors. The photosynthetic machinery intercepts solar radiation and converts CO_2 into energy in the form of sugars that can be directed towards dry matter accumulation. Therefore, understanding the different aspects of photosynthesis regulation and establishment, together with partitioning of assimilated carbon (C) to harvested organs, are examples of key mechanisms for increasing plant yield.

Sugarcane is one of the species with higher biomass production [39,40], due its high radiation use efficiency (RUE), high tillering, and leaf biomass. However, RUE is not constant throughout the plant lifecycle (for a review, see [41]), reducing its growth even when climatic and nutrition conditions are ideal, turning final production below the theoretical potential.

One of the most studied causes of the decrease in productivity is the reduction of leaf nitrogen (N) content [41], which is essential for maintaining high photosynthetic rates [42]. Aside from being a component of nucleic acids, proteins, and enzymes, N is a promoter of suckering [43, 44, cited in 2] and tillering [45, 46, cited in 2]. Since N is one of the main nutrients found in treated sewage effluents, its use could be a viable alternative that results in increased photosynthetic rates. However, as a plant ages, new completely expanded leaves lose their photosynthetic capacity, and this diminishing is independent of supplementation with N [47, 48]. Furthermore, self-shading has also been reported as a potential limitation on total plant photosynthetic capacity [49, 50].

A constituent of ATP, ADP, and several other molecules such as nucleic acids, phosphorus (P) is crucial to cell division, the growth of root systems, tillering, and shoots. However, only a small amount of the P content in the soil is available for plant uptake. Therefore, P is a nutrient to be evaluated in field experiments that intend to increase the photosynthetic rate. Potassium is also one of the several nutrients required for photosynthesis, starch synthesis, osmoregulation, and the uptake of water and stimulation of phloem transport [2].

There is a huge variation in photosynthetic capacity among sugarcane varieties [49, 51, 52], and despite the intense breeding and agronomic efforts to obtain more productive varieties, there is no clear correlation between an increase in photosynthetic efficiency and productivity [53]. Some reports indicate that sugarcane can increase its photosynthetic capacity when growing under elevated atmospheric CO_2 with consequent increment in productivity [54 - 56]. However, the evaluation of photosynthetic behavior and its correlation with the final biomass has been used as a prediction parameter during stressful conditions [57 - 63].

Crops possessing the C4 type of photosynthesis are characterized by the specialization of two cell types (bundle sheath and mesophyll, arranged in Kranz anatomy) working in concert to maximize CO_2 capture and fixation and simultaneously diminishing CO_2 leakiness and photorespiration [64 - 66]. These plants are among the most efficient biomass producers on earth, and some species, mainly grasses such as sugarcane, are used as feedstock for bioenergy production.

The modern sugarcane varieties are interspecific polyaneuploid hybrids between the primitive *Saccharum officinarum* and *S. spontaneum*, known for their capacity for accumulating high sucrose concentrations inside the vacuoles of parenchyma cells in the internodal culm [67]. Sucrose is the main product of photosynthesis, and is the transported sugar and the storage reserve carbohydrate in sugarcane. The first stage of photosynthesis, also known as light-dependent reactions, uses sunlight to trigger electron transport throughout several carries on the chloroplast thylakoid membranes, and it culminates in the production of reduction power in the form of NAPH and ATP and the release of oxygen to the atmosphere. These two products are then used in light-independent reactions to fixate CO_2 into carbohydrates through the Calvin-Benson-Bassham cycle inside chloroplast stroma. Sucrose is loaded into the conducting cells of the phloem (known as the sieve elements) for transport around the plant, and is unloaded in storage (stems) or growing sink organs (meristems, growing roots, and developing leaves). There, sucrose is processed as a source of C for cell growth [68].

Sucrose metabolism involves various enzymes and their isoforms. Sucrose synthesis is restricted to cytosol and can occur *via* two different routes. The first involves the enzyme sucrose-phosphate synthase (SPS) that catalyzes the transfer of the glucosyl unit from UDP glucose to fructose-6-phosphate, producing sucrose-6^F-phosphate (Suc6P) and UDP. The second, sucrose-phosphatase (SPP), hydrolyzes Suc6P to form sucrose [69]. The first reaction is reversible [70], but when sucrose-phosphatase hydrolyzates sucrose-6^F-phosphate, it is irreversible [71]. Alternatively, sucrose synthase (SuSy) catalyzes the reversible conversion of UDP glucose and fructose into sucrose and UDP. Both SPS and SuSy are able to synthesize sucrose, and the contribution of each enzyme is dependent on the tissue (leaf or culm) and developmental stage [72]. However, it is widely accepted that SPS plays a more prominent role in sucrose synthesis [68, 73 - 76] and is even considered as a biochemical marker for the high sucrose content in sugarcane [77].

Sucrose accumulation in the stems is a developmentally regulated process and starts after the early vegetative steps of stem growth and elongation. Sucrose accumulates at an increasing rate related to internode maturity and is characterized by an unceasing sequence of synthesis and degradation processes governed by several enzymes, known as a "futile cycle" due to the consumption of a considerable amount of energy in terms of ATP [76,78,79]. However, the fast sucrose processing in sink tissues is a strategy that allows plasticity due to changes in this sugar supply and demand at the whole-plant level [80]. In addition, this "futile cycle" is reported to enhance sink strength, reducing sucrose limitation of photosynthesis [81,82].

The activity of many sucrose-metabolizing enzymes changes according to the developmental stage of the culms. Both SPS and SuSy contribute to sucrose synthesis in young internodes, but SPS activity is higher than SuSy in mature internodes [76, 83]. Several studies report that the activity of the enzymes soluble acid invertase (SAI) [84 - 87], neutral invertase (NI) [88, 89], and cell wall invertase (CWI) [90] are high in immature internodes and decrease with the onset of culm maturation, but contradictory results have also been reported [91, 92]. It is believed that SAI and NI contribute to sucrose breakdown and the supply of hexoses to young growing tissues [89, 93], whereas CWI is associated with apoplasmic and symplasmic sucrose mobility [85, 94], playing a key role in phloem unloading [95].

The final sucrose concentration seems to be controlled by differences between the activities of SAI and SPS [86, 96 - 98]. Consequently, sucrose synthesis must exceed sucrose degradation for net sucrose accumulation to occur.

It is important to mention that total biomass, not only sugar content, is being focused on for renewable energy production [99]. Botha [100] compared the heat generated after combustion of sucrose and fiber, indicating that improved fiber content would be more beneficial even without taking into account the increase in photosynthesis and the advance in technologies for fermentation of ligno-cellulosic materials.

Breeding programs have negatively selected varieties with high fiber content [101], and as biomass total production has become more attractive, the low sucrose species *S. spontaneum* and its interspecific crosses have been more deeply characterized [102]. However, the physiological basis for the high fiber content of the so-called "energy canes" is still far from being elucidated. The increase in photosynthetic capacity has always been assumed to be limited by the negative feedback control by sugars [103]. However, Inman-Bamber and coworkers [99] studied clones with high fiber or sugar content and pinpointed that photosynthesis does diminish with the crop aging process, although no evidence was found that sugar content caused the decline in photosynthesis. The interspecific hybrid of *Saccharum* spp. L79–1002 has a high fiber content, and when compared to other C4 grasses its higher biomass production has been attributed to its longer growth rather than to higher radiation interception [99]. Related to E2G production, energy cane enables the utilization of the whole plant biomass above soil, which can double first generation production [104]. An intermediate approach has been used in breeding programs: that of producing an energy cane variety with high fiber content but also with high sugar. These two characteristics together enable the cane to be processed in existing mills [2].

Sugarcane bagasse has been extensively investigated as a source of sugars for producing E2G. Mostly composed of cell wall, the bagasse is usually burned to generate electric power in the vast majority of the mills. The use of bagasse to produce E2G is not yet economically viable, mainly because of sugarcane cell wall recalcitrance. This recalcitrance is principally due to limited access to the sugars inside cell wall components. Thus enzymes are used to degrade these limiting structures, in order to provide more accessible sugars that will be subsequently fermented.

The plant cell wall is a complex matrix of carbohydrate polymers, proteins, and other components. Plant cell walls act in cell growth and elongation [105], defense against pathogens [106], and provide structural support to cells, tissues, and the whole plant. In sugarcane, the cell wall also plays a role in sucrose transport [107]. Carbohydrates are the major components of cell walls, comprising about 90% of the cell wall mass. On the other hand, structural proteins account for 10% for dicots [108] and 1% for monocots [109]. Even though they present a small proportion, cell wall proteins comprise several hundred molecules playing different roles [110]. They can alter wall carbohydrate structure, play signaling roles, and interfere in the interactions with plasma membrane proteins on the cell surface [111]. Sugarcane cell wall proteins have been characterized [112], and this information can aid genetic engineering targeting altered cell walls. In turn, the polysaccharides that compose the plant cell wall may function as interconnected fibers or matrix, and are classified into pectins, hemicelluloses, and cellulose [113]. The synthesis of cellulose microfibrils coated with xyloglucans is accomplished by a complex of proteins that resides in the plasma membrane [2], playing an important role in wall structure [114]. Cellulose is the toughest and also the most stable cell wall polysaccharide. It contains continuous residues of β-1,4-glucopyranosyl, a type of binding that requires alternate residues to be placed 180° relative to each other. Usually, cell walls are about 30-50% cellulose, however, specialized cells can reach 95% cellulose, *i.e.,* cotton fibers

[115]. Because of this structure, cellulose has high tensile strength, which enables it to resist enzymatic degradation [2].

Monocots from the grass family, such as sugarcane, present differences regarding the presence of polysaccharides and proteins in comparison to dicots. The cell walls of grasses present cellulose fibrils interlocked by glucuronoarabinoxylans, differently from dicots which possess xyloglucans. Grasses also have a substantial portion of polymers other than cellulose connected to the fibrils by phenolic linkages resistant to alkali [116]. In addition, sugarcane cell walls present less pectin compared with dicots [117].

Lignin is another cell wall component, especially in secondary walls, and comprises around one third of the organic C from plant material [118]. However, lignin impairs enzymatic degradation [119]. The use of technologies to dismantle the plant cell wall requires a deeper understanding of cell wall structure and physiology. By using genetic engineering, it is possible to reduce lignin content, increasing cell wall digestibility without compromising plant development substantially [2]. The comprehension of the mechanisms involved in plant cell growth and development, especially as found in the cell wall, can therefore enable the modification of several characteristics related to E2G.

The partitioning of metabolites in response to the environment into growth, defense, and storage compounds, must be tightly regulated to promote biomass accumulation [120, 121]. Therefore, deciphering the relationship between metabolism and biomass accumulation can be considered a powerful tool for enhancing energy plant breeding programs [122]. In addition to biomass, relevant characteristics such as stress response, postharvesting, and nutrition depend on metabolic content [123].

Analysis of metabolic traits is valuable for enhancing the comprehension of the genes involved in the phenotypic variation related to metabolism [124]. It has been widely reported that studies of metabolic characteristics in the model plant Arabidopsis have increased knowledge of the issues concerning the control of metabolism and its impact on development and growth [121, 125 - 129]. One example is the untargeted metabolome analysis of an *Arabidopsis* recombinant inbred lines (RIL) population, which showed that around a third of the analytes have not been found in the parents, making room for plant breeding by altering metabolic patterns [125]. Recently, biomass accumulation and metabolic content have been addressed using *Arabidopsis* RILs populations [121, 126, 128, 130]. Results revealed that the correlation is not found with a single analyte, but between metabolic content and biomass, leading to the discovery of a metabolic signature for relevant characteristics [121]. These data were expanded to another analysis of the same RIL and near isogenic lines (NILs) populations, which revealed a couple of hotspots in which yield QTL overlapped with a large number of metabolite accumulation QTL [126]. A comparative metabolomic analysis of parental maize inbred lines and the hybrids exhibited distinct metabolic profiles [131]. Together, these data suggest that metabolite levels are subject to tighter control, since hybrids showed much more dense metabolic networks in comparison to inbred lines [131].

Due to the power of metabolic composition to predict yield, it is reasonable to speculate that this approach can be a powerful tool for sugarcane breeding. This is because it is not dependent on the genome sequence or on the comprehension of the segregation patterning from the progenies, as in the case of genetic markers. Similarly, data integration analysis (*i.e.,* metabolic profile data along with genetic markers, QTLs data, and phenotypic data) will provide further elucidation on gene function, something that is still very limited for sugarcane. This novel concept of plant performance prediction based on biomarker identification has great potential for supporting sugarcane breeding programs.

CONCLUDING REMARKS

Sugarcane ethanol production in Brazil has reached a worldwide status, both because of the energy content in this species and the future prospects for the use of renewable fuels. Traditionally, efforts inside the industry have been made toward increasing the final ethanol yield. Regarding plant breeding, work was focused on more productive varieties, adapted to different conditions. The use of agricultural management strategies, plant physiology knowledge, and innovative techniques for assisting the production of new varieties, can improve field and industry yield. Agricultural measures that increase the efficiency of water and nutrient uptake, together with varieties that are more productive in regard to both sucrose accumulation and fiber content, should enhance industrial ethanol production.

CONFLICT OF INTEREST

The authors confirm that this chapter content has no conflict of interest.

ACKNOWLEDGEMENTS

None declared.

REFERENCES

[1]   Instituto Brasileiro de Geografia e Estatística (IBGE). Municipal Agricultural Production , [12th Nov 2015]; Available at: http://www.sidra.ibge.gov.br/

[2]   Moore PH, Botha FC. Sugarcane: Physiology, Biochemistry, and Functional Biology. 1st ed., Oxford: John Wiley & Sons 2014.

[3]   Soares PA, Rossel CE. NAIPPE: o setor sucroalcooleiro e o domínio tecnológico. Rio de Janeiro: Nova Série 2007.

[4]   Instituto de Economia Aplicada (IPEADATA). [10th Dec 2015]; Available at: http://www. ipeadata.gov.br/

[5]   ÚNICA, Sugarcane Industry Association. O setor sucroenegético , [15th Feb 2015]; Available at: http://www.unica.com.br/

[6]   Scarpare FV, Leal MR, Victoria RL. Sugarcane ethanol in Brazil: challenges past, present and future. In: Dallemand JF, Hilbert JA, Monforti F, Eds. Bioenergy and Latin America: A Multi-Country Perspective JRC Technical Reports. Luxembourg: Publications Office of the European Union 2015; pp. 91-104.

[7]   Cardoso TF, Cavalett O, Chagas MF, *et al.* Technical and economic assessment of trash recovery in the sugarcane bioenergy production system. Sci Agric 2013; 70: 353-60.

[8]   Leal MR, Galdos MV, Scarpare FV, Seabra JE, Walter A, Oliveira C. Sugarcane straw availability, quality, recovery and energy use: A literature review. Biomass Bioenergy 2013; 53: 11-9.

[9]   Trivelin PC, Franco HC, Otto R, *et al.* Impact of sugarcane trash on fertilizer requirements for São Paulo, Brazil. Sci Agric 2013; 70: 345-52.

[10]   Dourado-Neto D, Timm LC, Oliveira JC, *et al.* State-space approach for the analysis of soil water content and temperature in a sugarcane crop. Sci Agric 1999; 56: 1215-21.

[11]   Graham MH, Haynes RJ, Meyer JH. Changes in soil chemistry and aggregate stability induced by fertilizer applications, burning and trash retention on a long-term sugarcane experiment in South African. Eur J Soil Sci 2002; 53: 589-98.

[12]   Cerri CC, Galdos MV, Maia SM, *et al.* Effect of sugarcane harvesting systems on soil carbon stocks in Brazil: an examination of existing data. Eur J Soil Sci 2011; 62: 23-8.

[13]   Figueiredo EB, La Scala N. Greenhouse gas balance due to the conversion of sugarcane areas from burned to green harvest in Brazil. Agric Ecosyst Environ 2011; 141: 77-85.

[14]   Ball-Coelho B, Tiessen H, Stewart JW, Salcedo IH, Sampaio EV. Residue management effects on sugarcane yield and soil properties in northeastern Brazil. Agron J 1993; 85: 1004-8.

[15]   Chapman LS, Larsen PL, Jackson J. Trash conservation increases cane yield in the Mackay district. Proc S Afr Sug Technol. 176-84.

[16]   Tominaga TT, Cássaro FA, Bacchi OO, Reichardt K, Oliveira JC, Timm LC. Variability of soil water content and bulk density in a sugarcane field. Aust J Soil Res 2002; 40: 604-14.

[17]   Andraski BJ, Mueller DH, Daniel TC. Effects of tillage and rainfall simulation date on water and soil losses. Soil Sci Soc Am J 1985; 49: 1512-7.

[18]   Timm LC, Oliveira JC, Tominaga TT, Cássaro FA, Reichardt K, Bacchi OO. Water balance of a sugarcane crop: quantitative and qualitative aspects of its measurement. Rev Bras Eng Agric Ambient 2002; 6: 57-62.

[19]   Fiorio PR, Demattê JA, Sparovek G. Cronologia e impacto ambiental do uso da terra na Microbacia Hidrográfica do Ceveiro, em Piracicaba, SP. Pesquisa Agropecu Bras 2000; 35: 671-9.

[20]   Macedo IC. A energia da cana-de-açúcar - doze estudos sobre a agroindústria da cana-de-açúcar no Brasil e sua sustentabilidade. São Paulo: Editor Berlendis & Vertecchia 2005; p. 245.

[21]   La Scala N Jr, Bolonhezi D, Pereira GT. Short-term soil CO_2 emission after conventional and reduced tillage of a no-till sugarcane area in southern Brazil. Soil Tillage Res 2006; 91: 244-8.

[22]   Braunack MV, McGarry D. Is all that tillage necessary? Aust Sugarcane 1998; 1: 12-4.

[23]   Segnini A, Carvalho JL, Bolonhezi D, *et al.* Carbon stocks and humification index of organic matter affected by sugarcane straw and soil management. Sci Agric 2013; 70: 321-6.

[24]   Scarpare FV, Hernandes TA, Ruiz-Corrêa ST, *et al.* Sugarcane water footprint under different management practices in Brazil: Tietê/Jacaré watershed assessment. J Clean Prod 2015; 1-9.

[25]   Food and Agriculture Organization of the United Nations (FAO). 2013. Available at: http://www.fao.org/nr/water/aquastat/countries_regions/brazil/index.stm

[26]   Scarpare FV. Bioenergy and water, bioenergy & water: Brazilian sugarcane ethanol. In: Dallemand JF, Gerbens-Leenes PW, Eds. Bioenergy and Water JRC Technical Reports. Luxembourg: Publications Office of the European Union 2013; pp. 89-101.

[27]   Gopinathan MC, Sudhakaran R. Biofuels: Opportunities and challenges in India. *In Vitro* Cell Dev Biol Plant 2009; 45: 350-71.

[28]   Instituto Brasileiro de Geografia e Estatística (IBGE). Agrarian census survey 2006 (Censo Agropecuário 2006) , 2006 [1st Aug 2013]; Available at: www.ibge.gov.br

[29]   Siebert S, Burke J, Faures JM, *et al.* Groundwater use for irrigation - a global inventory. Hydrol Earth Syst Sci 2010; 14: 1863-80.

[30]   Moreira JR. Water use and impacts due ethanol production in Brazil. International Conference on Linkages in Energy and Water Use in Agriculture in Developing Countries, Organized by IWMI and FAO, ICRISAT, India 2007. Available at: http://www.iwmi.cgiar.org/EWMA/files/papers/Jose_Moreira.pdf

[31]   Agência Nacional de Águas (ANA). Manual de conservação e reuso de água na agroindústria sucroenergética 2009. Available at: : http://www.ana.gov.br

[32]   Companhia de Tecnologia de Saneamento Ambiental (CETESB). Technical Standard P4231 e vinasse: criteria and procedures for application in agricultural soil (Norma Técnica P4231 e Vinhaça: Critérios e Procedimentos para Aplicação no Solo Agrícola) , [12th Nov 2013]; Available at: http://www.cetesb.sp.gov.br/Tecnologia/camaras/P4_231.pdf

[33]   Blum J, Melfi AJ, Montes CR. Nutrição mineral da cana-de-açúcar irrigada com efluente de esgoto tratado, em área com aplicação de fosfogesso. Pesquisa Agropecu Bras 2012; 47: 593-602.

[34]   Deon MD, Gomes TM, Melfi AJ, Montes CR, Silva E. Produtividade e qualidade da cana-de-açúcar irrigada com efluente de estação de tratamento de esgoto. Pesquisa Agropecu Bras 2010; 45: 1149-56.

[35]   Leal RM, Firme LP, Montes CR, Melfi AJ, Piedade SM. Soil exchangeable cations, sugarcane production and nutrient uptake after wastewater irrigation. Sci Agric 2009; 66: 242-9.

[36]   Leal RM, Firme LP, Montes CR, *et al.* Carbon and nitrogen cycling in a tropical Brazilian soil cropped with sugarcane and irrigated with wastewater. Agric Water Manage 2010; 97: 271-6.

[37]   Bond WJ. Effluent irrigation: an environmental challenge for soil science. Aust J Soil Res 1998; 36: 543-55.

[38]   Fonseca AF, Herpin U, Paula AM, Victoria RL, Melfi AJ. Agricultural use of treated sewage effluents: agronomical-environmental implications and perspectives for Brazil. Sci Agric 2007; 64: 194-209.

[39]   Robertson M, Wood A, Muchow R. Growth of sugarcane under high input conditions in tropical Australia. I. Radiation use, biomass accumulation and partitioning. Field Crops Res 1996; 48: 11-25.

[40]   Muchow R, Evensen C, Osgood R, Robertson M. Yield accumulation in irrigated sugarcane. II. Utilization of intercepted radiation. Agron J 1997; 89: 646-52.

[41]   Van Heerden PD, Donaldson RA, Watt DA, Singels A. Biomass accumulation in sugarcane: unravelling the factors underpinning reduced growth phenomena. J Exp Bot 2010; 61: 2877-87.

[42]   Sinclair T, Horie T. Leaf nitrogen, photosynthesis, and crop radiation-use efficiency: a review. Crop Sci 1989; 29: 90-8.

[43]   Stanford G. Sugarcane quality and nitrogen fertilization. Hawaiian Planters Record 1963; 56: 289-333.

[44]   Salter B, Bonnet GD. High soil nitrate concentrations during autumn and winter increase suckering. Proc Aust Soc Sugar Cane Technol 2000; 22: 322-7.

[45]   Das UK. Nitrogen nutrition of sugarcane. Plant Physiol 1936; 11: 251-317.

[46]   Wood RA. Nitrogen fertilizer use for cane. Sugar J 1968; 52: 3-15.

[47]   Allison J, Williams H, Pammenter N. Effect of specific leaf nitrogen content on photosynthesis of sugarcane. Ann Appl Biol 1997; 131: 339-50.

[48]   Park S, Robertson M, Inman-Bamber N. Decline in the growth of a sugarcane crop with age under high input conditions. Field Crops Res 2005; 92: 305-20.

[49]   Marchiori PE, Ribeiro RV, da Silva L, Machado RS, Machado EC, Scarpari MS. Plant growth, canopy photosynthesis and light availability in three sugarcane varieties. Sugar Tech 2010; 12: 160-6.

[50]   Marchiori P, Machado E, Ribeiro V. Photosynthetic limitations imposed by self-shading in field-grown sugarcane varieties. Field Crops Res 2014; 155: 30-7.

[51]   Irvine JE. Relations of photosynthetic rates, leaf and canopy characters to sugarcane yield. Crop Sci 1975; 15: 671-6.

[52]   Machado RS, Ribeiro RV, Marchiori PE, *et al.* Biometric and physiological responses to water deficit in sugarcane at different phenological stages. Pesquisa Agropecu Bras 2009; 44: 1575-82.

[53]   Makino A. Photosynthesis, grain yield, and nitrogen utilization in rice and wheat. Plant Physiol 2011; 155: 125-9.

[54]   Vu J, Allen L Jr, Gesch R. Up-regulation of photosynthesis and sucrose metabolism enzymes in young expanding leaves of sugarcane under elevated CO_2. Plant Sci 2006; 171: 123-31.

[55]   De Souza AP, Gaspar M, Da Silva EA, *et al.* Elevated CO_2 increases photosynthesis, biomass and productivity, and modifies gene expression in sugarcane. Plant Cell Environ 2008; 31: 1116-27.

[56]   Vu JC, Allen LH. Growth at elevated CO_2 delays the adverse effects of drought stress on leaf photosynthesis of the C(4) sugarcane. J Plant Physiol 2009; 166: 107-16.

[57]   Silva M, Jifon J, Da Silva J, Sharma V. Use of physiological parameters as fast tools to screen for drought tolerance in sugarcane. Braz J Plant Physiol 2007; 19: 193-201.

[58]   Da Graça J, Rodrigues F, Farias J, Oliveira M, Hoffmann-Campo C, Zingaretti S. Physiological parameters in sugarcane cultivars submitted to water deficit. Braz J Plant Physiol 2010; 22: 189-97.

[59]   Silva EN, Ribeiro RV, Ferreira-Silva SL, Vieira SA, Ponte LF, Silveira JA. Coordinate changes in photosynthesis, sugar accumulation and antioxidative enzymes improve the performance of *Jatropha curcas* plants under drought stress. Biomass Bioenergy 2012; 45: 270-9.

[60]   Ribeiro RV, Machado RS, Machado EC, Machado D, Magalhaes JR, Landell MG. Revealing drought-resistance and productive patterns in sugarcane genotypes by evaluating both physiological responses and stalk yield. Exp Agric 2013; 49: 212-24.

[61]   Silva MD, Jifon JL, dos Santos CM, Jadoski CJ, da Silva JA. Photosynthetic Capacity and Water Use Efficiency in Sugarcane Genotypes Subject to Water Deficit During Early Growth Phase. Braz Arch Biol Technol 2013; 56: 735-48.

[62]   Medeiros CD, Neto J, Oliveira MT, *et al.* Photosynthesis, antioxidant activities and transcriptional responses in two sugarcane (*Saccharum officinarum* L.) cultivars under salt stress. Acta Physiol Plant 2014; 36: 447-59.

[63]   Silva MD, Jifon JL, Da Silva JA, Dos Santos CM, Sharma V. Relationships between physiological traits and productivity of sugarcane in response to water deficit. J Agric Sci 2014; 152: 104-18.

[64]   Sheen JC. ₄ gene expression. Annu Rev Plant Physiol 1999; 50: 187-217.

[65]   Leegood RC. C_4 photosynthesis: principles of CO_2 concentration and prospects for its introduction into C_3 plants. J Exp Bot 2002; 53: 581-90.

[66]   Eberhard S, Finazzi G, Wollman F-A. The dynamics of photosynthesis. Annu Rev Genet 2008; 42: 463-515.

[67]   Inman-Bamber NG, Bonnett GD, Spillman MF, Hewitt MH, Glassop D. Sucrose accumulation in sugarcane is influenced by temperature and genotype through the carbon source-sink balance. Crop Pasture Sci 2010; 61: 111-21.

[68]   Lunn JE. Sucrose metabolism. Encyclopedia of life sciences. Chichester: John Wiley & Sons, Ltd 2008.

[69]   Lunn JE, MacRae E. New complexities in the synthesis of sucrose. Curr Opin Plant Biol 2003; 6: 208-14.

[70]   Barber GA. The equilibrium of the reaction catalyzed by sucrose phosphate synthase. Plant Physiol 1985; 79: 1127-8.

[71]   Lunn JE, ap Rees T. Apparent equilibrium constant and mass-action ratio for sucrose-phosphate synthase from seeds of *Pisum sativum*. Biochem J 1990; 267: 739-43.

[72]   Moore P. Temporal and spatial regulation of sucrose accumulation in sugarcane stem. Aust J Plant Physiol 1995; 22: 69-80.

[73]   Wendler R, Veith R, Dancer J, Stitt M, Komor E. Sucrose storage in cell suspension cultures of *Saccharum* sp. (sugarcane) is regulated by a cycle of synthesis and degradation. Planta 1990; 183: 31-9.

[74]   Huber SC, Huber JL. Role of sucrose-phosphate synthase in sucrose metabolism in leaves. Plant Physiol 1992; 99: 1275-8.

[75]   Stitt M. Metabolic regulation of photosynthesis. In: Baker NR, Ed. Advances in photosynthesis and respiration, Photosynthesis and the environment. Dordrecht: Kluwer Academic Publishers 1996; pp. 151-90.

[76]   Botha FC, Black GK. Sucrose phosphate synthase activity during maturation of internodal tissue in sugarcane. Aust J Plant Physiol 2000; 27: 81-5.

[77]   Grof CP, Albertson PL, Bursle J, Perroux JM, Bonnett GD, Manners JM. Sucrose-phosphate synthase, a biochemical marker of high sucrose accumulation in sugarcane. Crop Sci 2007; 47: 1530-9.

[78]   Batta SK, Singh R. Sucrose metabolism in sugar cane grown under varying climatic conditions: synthesis and storage of sucrose in relation to the activities of sucrose synthase, sucrose-phosphate synthase and invertase. Phytochemistry 1986; 25: 2431-1.

[79]   Whittaker A, Botha FC. Carbon partitioning during sucrose accumulation in sugarcane intermodal tissue. Plant Physiol 1997; 115: 1651-9.

[80]   Wang J, Nayak S, Koch K, Ming R. Carbon partitioning in sugarcane (*Saccharum* species). Front Plant Sci 2013; 4: 201.

[81]   Krapp A, Hofmann B, Schafer C, Stitt M. Regulation of the expression of rbcS and other photosynthetic genes by carbohydrates - a mechanism for the sink regulation of photosynthesis. Plant J 1993; 3: 817-28.

[82]   Koch KE. Sucrose metabolism: regulatory mechanisms and pivotal roles in sugar sensing and plant development. Curr Opin Plant Biol 2004; 7: 235-46.

[83]   Schäfer WE, Rohwer JM, Botha FC. A kinetic study of sugarcane sucrose synthase. Eur J Biochem 2014; 271: 3971-7.

[84]   Hatch MD, Glasziou KT. Sugar accumulation cycle in sugarcane. II. Relationship of invertase activity to sugar content and growth rate in storage tissue of plants grown in controlled environments. Plant Physiol 1963; 38: 344-8.

[85]   Gayler KR, Glasziou KT. Physiological functions of acid and neutral invertases in growth and sugar storage in sugarcane. Physiol Plant 1971; 27: 25-31.

[86]   Zhu YJ, Komor E, Moore PH. Sucrose accumulation in the sugarcane stem is regulated by the difference between the activities of soluble acid invertase and sucrose phosphate synthase. Plant Physiol 1997; 115: 609-16.

[87]   Rae AL, Casu RE, Perroux JM, Jackson MA, Grof CP. A soluble acid invertase is directed to the vacuole by a signal anchor mechanism. J Plant Physiol 2011; 168: 983-9.

[88]   Venkataramana S, Naidu KM, Singh S. Invertases and growth factors dependent sucrose accumulation in sugarcane. Plant Sci 1991; 74: 65-72.

[89]   Rose S, Botha FC. Distribution patterns of neutral invertase and sugar content in sugarcane internodal tissues. Plant Physiol Biochem 2000; 38: 819-24.

[90]   Lingle SE. Effect of transient temperature change on sucrose metabolism in sugarcane internodes. J Soc Sugar Cane Technologists 2004; 24: 132-41.

[91]   Dendsay JP, Singh P, Dhawan AK, Sehtiya HL. Activities of internodal invertases during maturation of sugarcane stalks. Int Sugarcane J 1995; 6: 17-9.

[92]   Vorster DJ, Botha FC. Sugarcane internodal invertases and tissue maturity. J Plant Physiol 1999; 155: 470-6.

[93]   Singh U, Kanwar RS. Enzymes in ripening of sugarcane at low temperatures. Int Sugarcane J 1991; 4: 2-4.

[94]   Sacher JA, Hatch MD, Glasziou KT. Sugar accumulation cycle in sugarcane. III. Physical and metabolic aspects in immature storage tissues. Plant Physiol 1963; 38: 348-54.

[95]   Roitsch T, Balibrea ME, Hofmann M, Proels R, Sinha AK. Extracellular invertase: key metabolic enzyme and PR protein. J Exp Bot 2003; 54: 513-24.

[96]   Ebrahim MK, Zingsheim O, El-Shourbagy MN, Moore PH, Komor E. Growth and sugar storage in sugarcane grown at temperatures below and above optimum. J Plant Physiol 1998; 153: 593-602.

[97]   Grof CP, Knight DP, McNeil SD, Lunn JE, Campbell JA. A modified assay method shows leaf sucrose-phosphate synthase activity is correlated with leaf sucrose content across a range of sugarcane varieties. Aust J Plant Physiol 1998; 25: 499-502.

[98]   Lingle SE. Sugar metabolism during growth and development in sugarcane internodes. Crop Sci 1999; 39: 480-6.

[99]   Inman-Bamber NG, Jackson PA, Hewitt M. Sucrose accumulation in sugarcane stalks does not limit photosynthesis and biomass production. Crop Pasture Sci 2011; 62: 848-58.

[100] Botha FC. Energy yield and cost in a sugarcane biomass system. Proc Aust Soc Sugar Cane Technol 2009; 31: 1-10.

[101] Wei X, Jackson P, Stringer J, Cox M. Relative economic genetic value (rEGV) – an improved selection index to replace net merit grade (NMG) in the Australian sugarcane variety improvement program. Proc Aust Soc Sugar Cane Technol 2008; 30: 174-81.

[102] Leal M. The potential of sugarcane as an energy source. Proc Aust Soc Sugar Cane Technol 2007; 26: 23-34.

[103] McCormick AJ, Watt DA, Cramer MD. Supply and demand: sink regulation of sugar accumulation in sugarcane. J Exp Bot 2009; 60: 357-64.

[104] Alexander AG. The energy cane alternative. Amsterdam: Elsevier 1985.

[105] Fry SC, Smith RC, Renwick KF, Martin DJ, Hodge SK, Matthews KJ. Xyloglucan endotransglycosylase, a new wall-loosening enzyme-activity from plants. Biochem J 1992; 282: 821-8.

[106] Bradley DJ, Kjelbom P, Lamb CJ. Elicitor-induced and wound-induced oxidative cross-linking of a proline-rich plant-cell wall protein - a novel, rapid, defense response. Cell 1992; 70: 21-30.

[107] Glasziou KT, Gayler KR. Sugar accumulation in sugarcane-role of cell walls in sucrose transport. Plant Physiol 1972; 49: 912-21.

[108] Cassab G, Varner JE. Cell wall proteins. Annu Rev Plant Physiol 1988; 39: 321-53.

[109] Vogel J. Unique aspects of the grasses cell wall. Curr Opin Plant Biol 2008; 11(3): 301-7.

[110] Carpita N, Tierney M, Campbell M. Molecular biology of the plant cell wall: searching for the genes that define structure, architecture and dynamics. Plant Mol Biol 2001; 47: 1-5.

[111] Jamet E, Canut H, Boudart G, Pont-Lezica RF. Cell wall proteins: a new insight through proteomics. Trends Plant Sci 2006; 11: 33-9.

[112] Calderan-Rodrigues MJ, Jamet E, Bonassi MB, *et al.* Cell wall proteomics of sugarcane cell suspension culture. Proteomics 2014; 14: 738-49.

[113] Carpita NC, McCann M. The Cell Wall. Biochemistry & Molecular Biology of Plants. Rockville: Am Soc Plant Physiol 2000.

[114] Arioli T, Peng L, Betzner AS, *et al.* Molecular analysis of cellulose biosynthesis in *Arabidopsis*. Science 1998; 279(5351): 717-20.

[115] Albersheim P, Darvill A, Roberts K, Sederoff R, Staehelin A. Plant Cell Walls. New York: Garland Science 2011.

[116] Carpita NC, Gibeaut DM. Structural models of primary-cell walls in flowering plants - consistency of molecular-structure with the physical-properties of the walls during growth. Plant J 1993; 1: 1-30.

[117] De Souza AP, Leite DC, Pattahil S, *et al.* Composition and structure of sugarcane cell wall polysaccharides: implications for second generation bioethanol. Bioenerg Res 2012.

[118] Pauly M, Keegstra K. Cell-wall carbohydrates and their modification as a resource for biofuels. Plant J 2008; 54: 559-68.

[119] Lu F, Ralph J. Lignin. In: Sun R-C, Ed. Cereal Straw as a Resource for Sustainable Biomaterials and Biofuels. 169-207.

[120] Tonsor SJ, Alonso-Blanco C, Koornneef M. Gene function beyond the single trait: natural variation, gene effects, and evolutionary ecology in *Arabidopsis thaliana*. Plant Cell Environ 2005; 28(1): 2-20.

[121] Meyer CR, Steinfath M, Lisec L, *et al.* The metabolic signature related to high plant growth rate in *Arabidopsis thaliana*. Proc Natl Acad Sci USA 2007; 104: 4759-64.

[122] Hirai MY, Klein M, Fujikawa Y, *et al.* Elucidation of gene-to-gene and metabolite-to-gene networks in *Arabidopsis* by integration of metabolomics and transcriptomics. J Biol Chem 2005; 280: 25590-5.

[123] Fernie AR, Schauer N. Metabolomics-assisted breeding: a viable option for crop improvement? Trends Genet 2009; 25: 39-48.

[124] Okazaki Y, Saito K. Recent advances of metabolomics in plant biotechnology. Plant Biotechnol Rep 2012; 6: 1-15.

[125] Keurentjes JJ, Fu J, Ric de Vos CH, *et al.* The genetics of plant metabolism. Nat Genet 2006; 38: 842-9.

[126] Lisec J, Meyer RC, Steinfath M, *et al.* Identification of metabolic and biomass QTL in *Arabidopsis thaliana* in a parallel analysis of RIL and IL populations. Plant J 2008; 53: 960-72.

[127] Kliebenstein DJ. Advancing genetic theory and application by metabolic quantitative trait loci analysis. Plant Cell 2009; 21: 1637-46.

[128] Lisec J, Steinfath M, Meyer RC, *et al.* Identification of heterotic metabolite QTL in *Arabidopsis thaliana* RIL and IL populations. Plant J 2009; 59: 777-88.

[129] Sulpice R, Pyl ET, Ishihara H, *et al.* Starch as a major integrator in the regulation of plant growth. Proc Natl Acad Sci USA 2009; 106: 10348-53.

[130] Meyer RC, Kusterer B, Lisec J, *et al.* QTL analysis of early stage heterosis for biomass in *Arabidopsis*. Theor Appl Genet 2010; 120: 227-37.

[131] Lisec J, Römisch-Margl L, Nikoloski Z, *et al.* Corn hybrids display lower metabolite variability and complex metabolite inheritance patterns. Plant J 2011; 68: 326-36.

CHAPTER 9

Industrial Use of Yeast (*Saccharomyces cerevisiae*): Biotechnology in Ethanol Biofuel Production Control in Brazil

Alessandro Antonio Orelli Junior[*]

Faculty of Technology of Piracicaba, Paula Souza Center, Piracicaba, Brazil

Abstract: The production of biofuels by fermentation is as old as mankind itself. The use of yeast (*Saccharomyces cerevisiae*) in the production of wine and beer, as well as bread, has been known from the time of the Rosetta Stone and the Bible. Nowadays, biotechnology is playing a major role in new advances in the fermentation of different substrates and in the production of ethanol, as well as in the production of biodiesel and many other biofuels. Advances in biotechnology have shed new light on ancient and well established technologies, bringing biofuels and mankind to a new, clean, sustainable, and bright future. A single and simple biotechnology method (electrophoretic karyotyping of intact chromosomes) has brought the entire ethanol industry in Brazil to new horizons.

Keywords: Bacteria contamination, Ethanol, Foam, Fingerprinting, Fructose, Glucose, Glycerol, Sucrose, Sugarcane, Yeast, Yield, Yeast monitoring.

INTRODUCTION

The production of biofuels by fermentation is as old as mankind itself. The use of yeast (*Saccharomyces cerevisiae*) in the production of wine and beer, as well as bread, has been known since the time of the Rosetta stone and the Bible. At present, biotechnology plays a major role in new advances in the fermentation of different substrates and in the production of ethanol, as well as in the production of biodiesel and many other biofuels. Advances in biotechnology have shed new light on ancient and well-established technologies, bringing biofuels and mankind to a new, clean, sustainable, and bright future.

"What cannot be measured cannot be managed." This maxim is particularly important when talking about biofuel production on a large scale, as in the case of the ethanol production sector in Brazil. Small increases in fermentation yields

[*] **Corresponding author Alessandro A. Orelli Jr.:** Faculty of Technology of Piracicaba, Paula Souza Center, Piracicaba-SP, Brazil; Tel: 55 19 982939540; E-mail: orelliaa@hotmail.com

represent large amounts of additional ethanol produced.

In the last thirty years, many techniques and protocols have been developed for understanding how to manage and improve yeast fermentation yields. Most of these techniques focus on the fermentation environment, but few are directly related to the yeast population itself.

The use of antibiotics to control bacterial infection is well known, but the infection of the fermentation by other "wild" yeasts, the ecology of the fermenting tank, and the relationship between different strains of yeast and between different strains of yeast and bacteria, are factors that are only now being brought to light by science.

THE ROLE OF YEAST (*SACCHAROMYCES CEREVISIAE*) IN ETHANOL FERMENTATION

The main physiological characteristic of yeast is its ability to degrade six-carbon carbohydrate (C6) molecules, such as glucose and fructose, to ethanol two-carbon (C2) components, without completely oxidizing them to CO_2, even in the presence of oxygen.

Yeast started to become a major ethanol producer about 80 million years ago, at the end of the Cretaceous period. A large amount of fruits were available, offering many fermentable substrates. Yeast developed the ability to accumulate and tolerate ethanol, inhibiting competing organisms from growing.

As a unicellular organism that clones itself, yeast has been "alive" at the surface of the earth ever since then, which indicates that it is very well adapted to the role of ethanol fermentation [1].

SUGARCANE ETHANOL PRODUCTION IN BRAZIL

In Brazil, the ethanol production industry is based on sugarcane and its large amounts of sugars (sucrose, glucose, and fructose) directly fermentable by yeast. The production of ethanol is conducted in sugar mills that produce only ethanol, or both ethanol and sugar. Each mill has a very different type of wort, depending on the availability of feedstocks such as sugarcane juice, sugarcane syrup, raw sugar, and molasses. The different amounts of each feedstock in wort production and the contamination of each component have direct implications for the population of bacteria and the different strains of yeast. The industry in Brazil does not sterilize the wort, and for this reason many different strains ("wild" yeast) are brought into the fermentation process. Any strain of yeast that is not selected or not known to the ethanol industry is commonly referred to as "wild" yeast.

"Wild" yeast is, 90% of the time, a bad strain for industrial ethanol fermentation due to foam production, low ethanol yields, residual sugar, etc. However, 10% of the time "wild" yeast can be a good new strain and thus receives attention from the industry and from academic experts.

THE FERMENTATION PROCESS IN THE SUGARCANE ETHANOL INDUSTRY IN BRAZIL

The main fermentation process used in Brazil is batch feed fermentation, with recirculation of the yeast. It consists of a very simple and efficient process.

The wort is produced using a mix of water, sugarcane juice, sugarcane syrup, raw sugar, and molasses (any of them in various proportions), reaching approximately 20% fermentable sugars.

The wort is then added to a fermenting vessel already containing treated yeast as 20% of its volume. In Brazil, there exist vessels of approximately one million liters.

The addition of the wort is done slowly in order to prevent the formation of foam and to allow the large yeast population to rapidly consume the sugar available in the wort, thus producing large amounts of ethanol. When the filling of the vessel is complete, two or three more hours are requited for the fermentation process to finish. The entire fermentation process totals about eight to twelve hours.

The so-called "wine" (fermented wort) is sent to centrifuges that will split the yeast cream from the wine. The yeast cream is sent to the treatment vessel, and the wine is sent to the distilling sector of the industry.

In the treatment vessel, the yeast receives water and sulfuric acid until reaching pH 2.5. The yeast stays for about two hours in this acid treatment, and is then sent back for a new fermentation.

This recirculation of yeast in the fermentation process is very important, because yeast consumes about 2 kg of sugars for each 1 kg of biomass produced. The loss of yeast biomass in the process reduces the ethanol yield.

"WILD" YEAST CONTAMINATION

The entire sugarcane ethanol process is conducted in a non-sterile environment and is directly affected by bacteria and "wild" yeast contamination. Bacteria are relatively easy to detect, and the use of antibiotics (natural or synthetic) controls the contamination level within reasonable parameters so as to prevent losses.

Formerly, in the 1980s, "wild" yeast was very difficult to detect. Plating with antibiotics in the culture medium selected the yeasts, but one could only see colonies with sharp or rugose formations, and brilliant or opaque colonies. The use of differential culture mediums was time-consuming and expensive. When the results were ready, it was too late to decide how to proceed with the process.

Today, in contrast, there are many techniques for fingerprinting different strains of yeast.

MONITORING THE YEAST

There is an evident relationship between the monitoring of the different populations of strains and factors such as fermentation, yields, productivity, stress, and consumption of acid.

The monitoring of yeast strains and ethanol production factors during the sugarcane harvest can also lead the industry to the best mix of strains for performing fermentation at a given moment of the harvest, in accordance with dry or rainy seasons, mix of sugarcane varieties being processed by the industry, and so on [2].

HOW TO MONITOR THE YEAST POPULATION: FINGERPRINTING

The most common fingerprinting methods are electrophoretic karyotyping of intact chromosomes and electrophoretic karyotyping of polymorphic fragments of mitochondrial DNA [3].

Electrophoretic karyotyping of Intact Chromosomes

The electrophoretic karyotyping of intact chromosomes is based on the separation of the chromosomal DNA from the yeast in an agarose gel. This method is based on the electrophoretic separation of the intact chromosomal DNA, molecules located in the cell nucleus that carry the genetic characteristics.

This method is excellent for differentiating species and different strains of the same species [4].

The method consists in plating samples and incubating them for approximately four days. After the growing of the yeast colonies, they are marked, and then cells are removed and fixed in agarose gel plugs. These cells will have the cellular walls, proteins, and lipoproteins associated with the DNA removed by enzymatic digestion. The plugs are washed many times, and then fixed in agarose gel and submerged in a buffer solution at 13°C. They are then subjected to pulsed field electrophoresis or transverse alternating field electrophoresis (TAFE). After eighteen hours the staining of the DNA is done, and the gel is photographed with ultraviolet light [5].

Electrophoretic karyotyping of Polymorphic Fragments of Mitochondrial DNA

This method consists in the extraction and digestion of the mitochondrial DNA with restriction enzymes and RNase, and in submitting the plugs to a similar process as that described for intact chromosomes [6].

This method is utilized not only to identify ethanol biofuel producing yeasts, but also to characterize the biodiversity of cachaça (cane spirits) producing yeasts [6].

The Industrial Yeast Fingerprint

The electrophoretic karyotyping of intact chromosomes is the most suitable technique for the monitoring and selection of yeast (*Saccharomyces cerevisiae*) in industry. This is because of the short time required for producing results compared to the other methods, its low DNA contamination risk, and its differentiation between *Saccharomyces* and non-*Saccharomyces* strains [6].

INDUSTRIAL LOSSES BY "WILD" YEAST

Nowadays, using any or all methods for fingerprinting yeast strains, we are able to monitor yeast (*Saccharomyces* and non-*Saccharomyces)* populations at any time of the harvest season and make decisions that will improve ethanol productivity (liters/day) and ethanol yield (liters/kilo of fermentable sugar).

The right choice of strain at the beginning of the harvest season is essential to maintaining high productivity and high yields. There are many selected industrial strains of yeast (*Saccharomyces cerevisiae*) available today, such as PE-2, CAT-1, BG-1, CR-1, FT-858, and so on. These yeasts should be dominant and persistent in the fermentation process during the entire harvest season.

DOMINANT AND PERSISTENT YEAST

A yeast strain is dominant when it has a small lag phase and multiplies itself in a very fast way, rapidly consuming the fermentable sugars and nutrients from the medium, producing large amounts of ethanol, and leaving other yeast strains behind.

A yeast strain is persistent when it remains in the fermentation during the harvest season for a long period of time, side by side with the "wild" yeast strains. These "wild" strains come from the sugarcane fields and are much more adapted to environmental conditions than are introduced strains.

FOAM

Rugose colonies of "wild" yeast notably produce a large amount of foam, consuming large amounts of dispersants and antifoaming products. The high levels of consumption of these products impacts directly in the cost of the ethanol produced.

FLOCCULATION

"Wild" yeasts are more susceptible to flocculation and decantation at the bottom of the fermentation vessel. This phenomenon decreases the surface contact of the yeast with the substrate and lowers the fermentation yield.

GLYCEROL

"Wild" yeasts are more susceptible to stresses like high temperature, high sugar and mineral salt content in wort, and stress caused by the presence of bacteria.

CONCLUSION

The use of the electrophoretic karyotyping of intact chromosomes has allowed the ethanol industry to monitor yeast strains active in fermentation at a reasonable cost and with fast response time. It is a decision tool that allows the ethanol industry to select the best strain or mix of strains for conducting wort fermentation at any time and under any conditions. Small increases in yields and productivity (lower foam formation, no flocculation, low glycerol production), obtained by the right use of the right yeast strain, have made possible the production of large amounts of ethanol and lowered the costs of production. Traditional methods of mass selection of yeast strains are leading the ethanol industry in Brazil to new levels of production yields and productivity.

CONFLICT OF INTEREST

The authors confirm that this chapter content has no conflict of interest.

ACKNOWLEDGEMENTS

None declared.

REFERENCES

[1]   Piskur J, Rozpedowska E, Polakova S, Merico A, Compagno C. How did Saccharomyces evolve to become a good brewer? Trends Genet 2006; 22: 183-6.

[2]   Basso LC, Amorim HV. Relatório Annual de Pesquisas em Fermentação Alcoolica Piracicaba: Fermentec 1993.

[3]   Amorim HV, Leão RM. Fermentação Alcoolica: Ciencia e Tecnologia Piracicaba: Fermentec 2005.

[4]   Querol A, Barrio E, Huerta T, Ramon DA. Molecular Monitoring of Wine fermentations Conducted by Dry Yeasts Strains. Appl Environ Microbiol 1992; 58: 2948-53.

[5]   Santiago LC, Oliveira Junior WP. Utilização da analise de DNA mitocondrial para caracterizar a biodiversidade de leveduras isoladas da fermentação da cachaça. Seminario de iniciação científica. Palmas-TO 2012.

[6]   Gomes LH. Avaliação de Quatro Metodos para a Caracterização de Leveduras. Tese ESALQ/USP Piracicaba-SP 1995.

CHAPTER 10

Biodiesel from Microalgae: Third Generation Biofuel

Gisele G. Bortoleto[*]**, Henrique L. de Miranda** and **Rodrigo H. de Campos**

Faculty of Technology of Piracicaba, Paula Souza Center, Piracicaba, Brazil

Abstract: At present, fuels derived from biomass play an important role in scenarios for the expansion of renewable energy worldwide. Considering the liquid biofuels, biodiesel is an interesting alternative that minimizes mineral diesel consumption. Among the raw materials available for biodiesel production, microalgae biomass has been described as an alternative with great potential for accomplishing the goal of replacing diesel by biodiesel without competing with fertile land for food production. This paper presents advances and challenges in the technologies presently used for the production of biodiesel from microalgae, including the procedures used to obtain biomass and the evolving technologies for reducing production steps (and consequently time and process costs). It was found that microalgae is currently still not a viable option for large-scale biodiesel production, as it has a negative energy balance. On the other hand, microalgae indicates substantial earning potential in biomass and lipid fractions, being a good alternative because of its high fat content. Therefore, research directed toward the production of biofuels from microalgae should receive greater attention and investment. Microalgae can become a competitive alternative and a commercial reality in the biofuels sector with the development of genetic improvement and technological production systems and with hoped-for reductions in costs, and also because of its sustainable qualities in contrast to other raw materials.

Keywords: Biofuel, Biodiesel, Biomass, Microalgae.

INTRODUCTION

The growing demand for energy on the world stage, predictions of high trends in oil prices, concerns about the environmental impact of the use of fossil fuels, energy security, and governmental and social incentives, all justify the prospects for expansion of renewable energy worldwide. Fuels derived from biomass play an important role in this scenario, and, among the available biofuels, biodiesel stands out among the most promising. It consists of alkyl esters of long chain carboxylic acids produced from the transesterification and/or esterification of raw

---

[*] **Corresponding author Gisele G. Bortoleto:** Faculty of Technology of Piracicaba, Paula Souza Center, Piracicaba-SP, Brazil; Tel: +55 19 97147 7444; E-mail: gisele.bortoleto@fatec.sp.gov.br

Daniela Defavari do Nascimento, & William A. Pickering (Eds.)

greases and fats of vegetable and animal origin. This biofuel is an attractive alternative to petroleum-derived liquid fuel and can be used without major changes in current engine design, generating sustainability in the sector by reducing greenhouse gas emissions. It can also be produced from different sources, such as vegetable oils (soybean, palm, sunflower, cotton, peanut, *etc.*), animal fat, and reused oil (fry oil).

Recent studies have also shown the potential of unusual oil sources, especially the production of oil from microalgae. Microalgae are a diverse group of prokaryotic and eukaryotic microorganisms with chlorophylls and/or other photosynthetic pigments, which have the ability to perform photosynthesis. Due to their simple structures and their efficient photosynthetic systems for conversion of solar energy into organic compounds, they have the ability to grow rapidly and store large amounts of lipids. In general, microalgae present themselves as an alternative for biodiesel production due to high productivity of lipid biomass in a very short production cycle. Microalgae also do not require fertile and/or arable land, and thus do not compete with food production and do not cause deforestation. Furthermore, they can even be used as mitigating sources of greenhouse gases from stationary sources. However, microalgae biomass production and the subsequent production of biodiesel consist of several stages toward attaining the final product, with various technical and economic barriers that need to be analyzed and controlled before the raw material becomes a viable alternative to large-scale production of biomass for energy purposes. The steps of the process are presented in this paper, showing the different methods and technological systems employed, as well as their challenges. The present paper analyzes a variety of studies on this topic, evaluating the use of microalgae for biodiesel production and studying its potential as a renewable energy source and its environmental and economic viability.

BIOFUELS

Biofuels are alternative biodegradable fuels produced from biomass (organic raw materials) derived from renewable energy sources. They have become a sustainable alternative to "traditional" fuels, as they are able to bring significant energy security and environmental benefits [1]. Their production has become a solution for reducing dependency on fossil fuels currently used in transportation vehicles and various other industrial processes [2]. Primary biofuels such as firewood, for example, are used without processing, mainly for heating, cooking, or electricity production. Secondary biofuels, such as bioethanol and biodiesel, are produced from the processing of biomass [3]. Secondary biofuels can be classified into four generations based on different parameters such as type of raw material, processing technology, and level of development. Third generation biofuels are produced from microalgae cultivation, where cells accumulate lipids which, after extraction, pass through the transesterification process for obtaining biodiesel [4 - 7].

BIODIESEL HISTORY

In 1895, about twenty-five years after the beginning of the oil industry, Rudolf Diesel created the compression ignition engine, which was later named in his honor [8]. In 1900, during the Paris Exposition, the French company Otto, at the request of the French government, presented a small diesel engine running on peanut oil. Peanuts were a crop that was widespread in French colonies in Africa and could be easily cultivated locally, thus meeting the energy demands of colonies and industries. In the 1940s, several European countries that had African colonies such as Belgium, France, Italy, the UK, and even Germany, showed interest in the production and development of fuel through vegetable oils. During World War II, vegetable oils were used as emergency fuel for various applications in countries like Brazil, Argentina, China, India, Japan, and the United States. At this time, concerns about excessive use of oil products, and about their possible scarcity, stimulated the development of alternatives such as mixtures (biofuels) using cottonseed oil, corn, and mixtures with conventional diesel [9]. However, at the end of the war, the return of a plenteous and low priced supply of imported oil, mainly from the Middle East, ended up discouraging the use, and thus the development, of alternative fuels.

Biodiesel in the World

Despite the decline suffered by the sector after the war, advances in the development of biofuels have been gradually expanding around the world. According to the United States Energy Information Administration (EIA) [10], world production of biodiesel increased steadily from 213 million gallons in 2000 to 6289 million gallons in 2013. The biodiesel market worldwide arose from the need for reducing the use of fossil diesel. In addition, the anthropogenic causes of global warming described by the Kyoto Protocol decreed the urgent need for new energy technology. Another variable is the unquestionable political and social instability in major oil producing countries [11, 12].

Developing countries have a great opportunity to rise in the biofuels market, and these emerging countries can stimulate this segment of their economies, the development of industrial parks, the use of technology, the exploration of new markets, and the forming of partnerships. According to BP Global, biodiesel has been widely used in the EU (mainly Germany and France) from 2014 on [13], using the rapeseed oil surplus and rapeseed. The alcohol used in Europe is methanol, which can be purchased at extremely competitive prices due to the installation of several factories in the Middle East. One oil among others used for the production of fuels is sunflower oil. Other countries that have been producing biodiesel in Europe are Italy, Spain, and Poland [14, 15].

In North America, especially in the United States, the biofuels program has been developing intensively. After the first oil crisis, research emphasized the production of ethanol from corn and, from the end of the 1990s onward, the production of biodiesel obtained mainly from soybean and rapeseed [14].

Despite the unquestionable quantitative difference in production and consumption of ethanol and biodiesel, the latter has been developing gradually, reaching an output of 1339 billion liters in 2013 [15]. The United States is the world's largest consumer of fossil fuel. To reduce this dependence on fossil fuels from the Middle East and Venezuela, incentives are being provided in the US for domestic biodiesel production. Malaysia has the world's largest palm production, and plans to export to the European Union. China has invested extensively in research for the improvement of biodiesel technology, since the high growth rate of its economy is linked to a large consumption of fossil fuels. However, the widespread adoption of biodiesel in China comes up against the barrier of the need for fields to produce food for its large population. Other countries that wish to follow the path of biofuels have not yet established policies regarding the adoption of new technologies or have no representative volume of biodiesel production or biodiesel consumption [12].

BIODIESEL PRODUCTION: TRANSESTERIFICATION REACTION

According Khalil [16], the sources of inputs for classical production of biodiesel are comprised of the following components: triglycerides, alcohol, and catalyst. Animal fats and vegetable oils are comprised of triglyceride molecules consisting of three long chain fatty acids, linked to a glycerol molecule in the form of esters. These fatty acids vary in the length of the carbon chain, and in the number, orientation, and position of the double bonds [17]. However, due to a number of limiting factors such as high viscosity, the presence of free fatty acids, incomplete combustion, and low volatility, the use of vegetable oils as an alternative fuel in their pure form has shown to be unsatisfactory and impractical, requiring derivatization of triglyceride to allow their use in diesel engines without problems [9, 17]. Various methods have been investigated to reduce these problems, for example, the use of binary mixtures with petroleum diesel, thermochemical process, micro-emulsification and transesterification, the latter being the only method that leads to the commonly denominated "biodiesel", or fatty acid alkyl ester [9]. In the transesterification process, one mole of triglyceride is reacted with three moles of alcohol, methanol, or ethanol, in the presence of a catalyst which can be homogeneous, heterogeneous, or enzymatic [18], yielding monoesters of fatty acids and glycerin as a byproduct [19]. Among the factors that influence transesterification are the following: the type of alcohol; the molar ratio of glycerides and alcohol; type and amount of catalyst; temperature; reaction time; content of free fatty acids (FFA); and water present in the grease raw material [20]. In the process of transesterification, alcohols are key substrates. The most commonly used alcohols are methanol, ethanol, propanol, butanol, and amyl alcohol. Methanol is used most widely because of its low cost, lower power consumption, higher reactivity allowing spontaneous glycerol separation, and higher performance and reduced moisture content as compared to ethanol. Transesterification using ethanol is more laborious, since the use of ethanol, even anhydrous ethanol, implies problems in glycerol separation from the reaction medium [21]. In Brazil, however, ethanol is an interesting option due to relevant factors such as supply issues, logistics, and operational safety [16].

The reaction proceeds as an equilibrium reaction and therefore requires the provision of excess alcohol for maintaining balance change for the product and for enhancing its rate of reaction [22]. High temperatures allow higher yields in shorter times, but it is necessary to assess whether the expenditure of the energy required for heating does not exceed the gains from the time savings. Vigorous stirring is another important aspect for achieving high yields, as homogeneity of mixing alcohol/vegetable oil is essential to efficiently proceed with transesterification. However, after the system mixing, vigorous stirring can cause scattering of glycerol droplets in the reaction medium. This phenomenon can result in a very slow coalescence of glycerol and, consequently, a greater time required for separation [21].

The transesterification reaction can be catalyzed by both homogeneous and heterogeneous catalysts, which can be acidic, basic, or enzymatic [21, 23]. Base homogeneous catalysis is the preferred method for industrial processes, because it has high yields and rates of reaction and conversion that are much higher than acid catalysis. However, the transesterification of vegetable oils in alkaline medium has the inconvenience of producing soaps, both by neutralization of free fatty acids and by saponification of the glycerides and/or monoalkyl esters that are formed. Such side reactions are undesirable because they consume part of the catalyst by decreasing the yield of the transesterification of glycerol and complicating the process of separation and purification of biodiesel, requiring the addition of more catalyst to compensate for the loss [21, 23]. When the level of free fatty acids (FFA) is greater than 5%, a pretreatment converting FFA to methyl esters for the reduction of FFA levels is required. With reduced levels of free fatty acids, the oil is then transesterified with an alkali catalyst [23]. The most widely used basic catalysts are sodium hydroxide (NaOH) or potassium hydroxide (KOH).

In the process of acid homogeneous catalysis, the most commonly used acids are hydrochloric acid (HCl), sulfuric acid (H_2SO_4), and anhydrous sulfonic acids. Transesterification using acid catalysts leads to high yields, but it has inconveniences such as a need for a high alcohol/oil molar ratio and requiring long periods of synthesis [21, 23].

In the heterogeneous catalysis process, enzymatic catalysis has been the most explored [21]. The process of enzyme catalysis, although it has a high tolerance for levels of free fatty acids (FFA), is expensive and cannot provide the degree of reaction necessary to meet biodiesel quality specifications [23].

Supercritical transesterification reactions (SC) do not require the use of catalysts for acceleration of the process [24]. This process has a high rate of reaction, simple separation, and purification of the final products, and no waste production related to the separation of the final product catalyst, making this an environmentally friendly process. However, due to the large amounts of capital needed to build supercritical reactors, and the high energy used in their operation, non-catalytic processes for biodiesel production do not yet have industrial applications [25].

RAW MATERIALS FOR BIODIESEL PRODUCTION

The most typical grease raw materials for biodiesel production are refined vegetable oils [26]. These inputs may be relatively expensive, even under the best of conditions, when compared to petroleum products. Therefore, the choice of raw materials varies from one location to the other according to their availability and corresponding economic viability. In addition to vegetable oils, other raw materials for biodiesel production are animal fats, residual oil, and fat [27]. Other inputs used in this sector are microalgae and macroalgae [22, 28].

Currently, over 95% of raw materials used in biodiesel production come from edible vegetable oils (first generation biofuels). These raw materials, although widely used, conflict with the food industry and can cause increases in food costs, thus impacting the sustainability of their production [22]. The production of second generation fuels, while not affecting food security and having significant advantages over first generation fuels, is currently unsustainable [22].

Faced with this problem, research has been done using microalgal biomass, considering it as an alternative in terms of environmental development, social and economic acceptability, and greater energy and food security. Despite the challenges encountered, researchers believe that the rewards will eventually outweigh the risks [7, 22, 29].

MICROALGAE FOR BIODIESEL PRODUCTION

Microalgae are a diverse group of prokaryotic and eukaryotic microorganisms that have chlorophylls and/or other photosynthetic pigments with the ability to perform photosynthesis. Examples of prokaryotic organisms are cyanobacteria (*Cyanophyta*); examples of eukaryotic microalgae are green algae (*Chlorophyta*) and diatoms (*Bacillariophyta*) [30, 31].

Microalgae are one of the oldest forms of life, and they are able to live in harsh conditions and in all ecosystems on the planet, not only in water but also on land. They represent a wide range of species living in various environmental conditions. Due to their simple structures and their efficient photosynthetic systems for conversion of solar energy into organic compounds, they have the ability to grow rapidly [28, 30, 31].

In microalgal biomass composition, the main component is carbon (C) from inorganic (CO_2, HCO_2^{-3}) and organic (pentoses, hexoses) sources. In addition to carbon (C), several micronutrients are present, such as Si, Fe, and Mn, that participate in primary reactions in the life of the cell. Macronutrients are also present, such as phosphorus (P) and nitrogen (N), which form the structural components of the bio cell, the cell membrane, and intracellular structures, and also participate in energy processes and metabolic regulation. Vitamins such as cobalamin (B12), biotin, and thiamine (B1) are compounds essential for the functioning of the metabolism, playing a role in maintenance and growth [32, 33]. For their growth, these phototrophic organisms need sunlight, carbon dioxide (CO_2), water, inorganic salts (N, P, K), and a temperature range of 20-30°C.

Microalgae can be autotrophic or heterotrophic, and the former use inorganic compounds as a carbon source. They can be photoautotrophic, using only light as an energy source, or chemoautotrophic, oxidizing inorganic compounds for energy. Heterotrophic microalgae use organic compounds as a source for energy and growth. These can be photoheterotrophic, depending upon organic compounds as a carbon source and using light for their development, or they can be chemoheterotrophic, obtaining energy through oxidation of organic compounds. Some microalgae are mixotrophic, combining heterotrophy and autotrophy by photosynthesis [5].

Microalgae can be used to mitigate greenhouse gases, as they have the ability to fix carbon dioxide (CO_2) from various sources such as the atmosphere, gases emitted by industry, and soluble carbonates [34]. Worldwide, several species of microalgae were analyzed and studied, and in them were found a wide range of high-value-added bioproducts with broad applications in health, nutrition, aquaculture, cosmetics, and the environment [7, 28, 31, 35]. Another relevant application that has aroused the interest of researchers is the use of microalgae as an energy alternative to fossil fuels [28, 31, 35, 38].

Microalgae Potential in Biodiesel Production

From the year 1970 on, the National Renewable Energy Laboratory (NREL) of the United States, funded by the U.S. Department of Energy (DOE), initiated a program focused on alternative renewable biofuel. Beginning in the early 1980s, the production of biodiesel from microalgae was included in this program [39].

The main objectives of this program were the following: the study of the biochemistry and physiology of the production of lipids in oil producing microalgae; concluding whether the use of microalgae for low cost production of biodiesel is viable; the production of improved varieties of algae; and the search and exploitation of genetic variability of algal viruses as potential genetic vectors.

Several favorable factors in using microalgae as a renewable source of biodiesel, such as the high efficiency of microalgae in solar energy storage, cell composition with high oil yield, fast growth, and non-competition with food, among others, sparked a wave of discussion and investment in research on the high-yield production of this renewable source as compared to other terrestrial crops [37, 38].

The photosynthetic potential of microalgae, with its CO_2 fixation capacity, provides an efficient microalgal photoconversion process, thus producing high yields of biomass and lipid fractions in controlled cultures. It was found that microalgae, when compared to other crops for biodiesel production, presents itself as a very good alternative because of its high oil content [38]. Although the quantities of oil produced by microalgae vary by type of crop and biochemical composition, the oil content can exceed 80% by weight of dry mass in some cases, and a range of 20-50% is commonly found [28]. Table **1** compares the efficiency of microalgae for biodiesel production and land use with other plant oil crops, and compares their oil content in dry biomass and their oil production per hectare per year [7].

Although the oil content of plants with seeds and microalgae are similar, there are significant variations in biomass productivity, which results in a great advantage for microalgae in oil yield and biodiesel productivity. As for land use, microalgae, followed by palm oil, are clearly advantageous due to their increased biomass productivity and oil yield.

Table 1. Comparison of microalgae with other raw materials for biodiesel production.

Plant oil source	Oil content (% oil by weight in biomass)	Oil yield (L oil/ha/ year)	Land use (m² year/kg biodiesel)	Biodiesel productivity (kg biodiesel/ ha/year)
Corn (*Zea mays* L.)	44	172	66	152
Hemp (*Cannabis sativa* L.)	33	363	31	321
Soy (*Glycine max* L.)	18	636	18	562
Jatropha (*Jatropha curcas* L.)	28	741	15	656
Camelina (*Camelina sativa* L.)	42	915	12	809
Canola/ Rapeseed (*Brassica napus* L.)	41	974	12	862
Sunflower (*Helianthus annuus* L.)	40	1070	11	946
Castor (*Ricinus communis*)	48	1307	9	1156
Palm oil (*Elaeis guineensis*)	36	5366	2	4,747
Microalgae (low oil content)	30	58,700	0,2	51,927
Microalgae (medium oil content)	50	97,800	0,1	86,515
Microalgae (high oil content)	70	136,900	0,1	121,104

In contrast to other alternatives, biomass derived from microalgae as a renewable source of biodiesel has several advantageous features, such as: (1) production does not depend on harvest times, so production is year round and harvest is multiple and continuous due to rapid growth, easily surpassing the performance of other oil crops; (2) despite being cultivated in an aqueous medium, microalgae requires less water than other terrestrial crops, and microalgae may occur in non-cropland areas and in brackish and waste water, reducing the need for the resources needed for the production of cultures; (3) there is no need to use chemicals such as pesticides and herbicides, which helps reduce costs and environmental impacts; (4) a large carbon dioxide fixation capacity (CO_2), contributing to the mitigation of greenhouse gas emissions; (5) the cultivation of microalgae in sewage is possible, using compounds such as nitrate, phosphate, phosphorus, and nitrogen as nutrients for microalga growth, and in addition acting as an agent in the treatment of these wastewaters; (6) after oil extraction, the residual biomass may contain high-value-added byproducts in the form of proteins, biopolymers, carbohydrates, and pigments, which can be used in the pharmaceutical or other commercial sectors; (7) different species of microalgae can be adapted to live in their respective environmental conditions of production, and the best species can be suited to environments with specific growth characteristics, which is not possible with other oilseed cultures [7, 22, 28, 35 - 38, 40 - 42].

These factors have attracted interest in development, research, and innovation, with the primary objective of overcoming the economic and environmental barriers to this new opportunity for the renewable energy sector [7].

Selection of Microalgae: Species and Strains

Among the microalgae there are several groups that differ in pigment composition, biochemical constituents, structure, and life cycle. Among them, some of the groups used in research on biodiesel production are diatoms (Class *Bacillariophyceae*), green algae (Class *Chlorophyceae*), chrysophytes (Class *Chrysophyceae*), prymnesiophytes (Class *Prymnesiophyceae*), eustigmatophyte (Class *Eustigmatophyceae*), and also cyanobacteria (Class *Cyanophyceae*) [39]. In order to use these bodies as feedstock for biofuels (and other products with commercial value), study and characterization is necessary for those species which can adjust their physiology to the environmental conditions of local cultivation. This characterization is a difficult task, as it requires the bioprospecting of a large number of species [43].

The study and characterization of species allows for the selection of species that meet the needs for the production of the final product with the highest possible productivity. The selection of species with high lipid content, as well as high productivity of microalgal biomass and especially high productivity of lipids (especially triacylglycerols), is the key to economically viable biodiesel production [22, 39].

Lipid productivity determines the cost of the cultivation process, while biomass productivity in culture and lipid content significantly affect the cost of the extraction and production processes. Therefore, the ideal process of biodiesel production should enable increased productivity of lipids and the highest biomass productivity [7]. Unfortunately, this ideal situation is very difficult to meet in practice because cells with high lipid content are produced under conditions of physiological stress, which is associated with conditions of nutrient limitation and, therefore, low biomass productivity [22, 44].

Most species of algae considered for biodiesel production are green algae (*Chlorophyta*) and diatoms (*Bacillariophyta*). They are usually photosynthetic, but several species are able to grow heterotrophically [39].

Not all microalgae lipids are suitable for biodiesel production, however, they are all appropriate for this purpose as fatty acids, free or covalent bonds to glycerol and its derivatives. They are produced frequently and constitute a larger fraction of the total lipids, generally between 20% and 40% [28]. For the development of quality biofuels, a long chain fatty acid profile with a low degree of unsaturation is desirable.

Table **2** presents the lipid content, lipid productivity, and biomass productivity of various species and strains of microalgae, both marine and freshwater, showing the differences among the various species and strains [7].

Table 2. Lipid content and productivity of different species/strains of microalgae.

Microalgal species	Lipid content (% dry mass concentration)	Lipid productivity (mg/L/day)	Volumetric biomass productivity (g/L/day)	Biomass productivity per unit area per unit time (g/m²/day)
Ankistrodesmus sp.	24.0-31.0	-	-	11.5-17.4
Botryococcus braunii	25.0-75.0	-	0.02	3.0
Chaetoceros muelleri	33.6	21.8	0.07	-
Chlorella emersonii	25.0-63.0	10.3-50.0	0.036-0.041	0.91-0.97
Chlorella protothecoides	14.6-57.8	12-14	2.0-7.7	-
Chlorella vulgaris	5.0-58.0	11.2-40.0	0.02-0.02	0.57-0.95
Chlorella sp.	10.0-48.0	42.1	0.02-2.5	1.61-16.47/25.0
Chlorella	18.0-57.0	18.7	-	3.50-13.90
Chlorococcum sp.	19.3	53.7	0.28	-
Crypthecoidinium cohnii	20.0-51.1	-	10.0	-
Dunaliella salina	6.0-25.0	116.0	0.22-0.34	1.6-3.5/20.0-38.0
Dunaliella tertiolecta	16.7-71.0	-	0.12	-
Dunaliella sp.	17.5-67.0	33.5	-	-
Ellipsoidion sp.	27.4	47.3	0.17	-
Euglena gracilis	14.0-20.0	-	7.7	-
Haematococcus pluvialis	25.0	-	0.05-0.06	10.2-36.4
Isochrysis galbana	7.0-40.0	-	0.32-1.6	-
Isochrysis sp.	7.1-33.0	37.8	0.08-0.17	-
Monodus subterraneus	16.0	30.4	0.19	-
Monallanthus salina	20.0-22.0	-	0.08	12.0
Nannochloropsis oculata.	22.7-29.7	84.0-142.0	0.37-0.48	-
Nannochloropsis sp.	12.0-53.0	37.6-90.0	0.17-1.43	1.9-5.3
Neochloris oleoabundans	29.0-65.0	90.0-134.0	-	-

(Table 2) contd.....

Microalgal species	Lipid content (% dry mass concentration)	Lipid productivity (mg/L/day)	Volumetric biomass productivity (g/L/day)	Biomass productivity per unit area per unit time (g/m²/day)
Nitzschia sp.	16.0-47.0			8.8-21.6
Oocystis pusilla	10.5	-	-	40.6-45.8
Pavlova salina	30.9	49.4	0.16	-
Pavlova lutheri	35.5	40.2	0.14	-
Phaeodactylum tricornutum	18.0-57.0	44.8	0.003-1.9	2.4-21
Porphyridium cruentum	9.0-18.8/60.7	34.8	0.36-1.5	25.0
Scenedesmus oblíquos	11.0/55.0	-	0.004-0.74	-
Scenedesmus sp.	19.6/21.1	40.8-53.9	0.03-0.26	2.43-13.52
Skeletonema costatum	13.5-51.3	17.4	0.08	-
Spirulina platensis	4.0-16.6	-	0.06-4.3	1.5-14.5/24.0-51.0
Spirulina máxima	4.0-9.0	-	0.21-0.25	25.0
Tetraselmis suecica	8.5-23.0	27.0-36.4	0.12-0.32	19.0
Tetraselmis sp.	12.6-14.7	43.4	0.3	-

As shown in Table **2**, although there are species that have a lipid content of 75% of the dry weight biomass, they do not exhibit good productivity. Species such as *Chlorella, Crypthecodinium, Cylindrotheca, Dunaliella, Isochrysis, Nannochloris, Nannochloropsis, Neochloris, Nitzschia, Phaeodactylum, Porphyridium Schizochytrium,* and *Tetraselmis*, despite having lower lipid content (between 20 and 50%), can achieve higher productivity.

Species such as *Chlorella* are suitable candidates for biodiesel production because they can produce about 63% of lipid content of dry weight biomass, besides having great flexibility to adapt to different environmental conditions [7, 44]. However, the selection of the most suitable species must take into account other factors, such as the ability of microalgae to develop using the available nutrients and other specific environmental conditions [7, 22, 39].

Screening of Species and Strains for Large-scale Production

The algae to be evaluated can come from samples collected in their natural environment, such as water or soil, or from already existing samples from pre-selected cultures. The samples collected in environments similar to the large-scale production environment are preferred [39 - 45].

The first step in microalgae screening is to enrich the samples with nutrients such as nitrogen (N) and phosphorus (P) up to the standard levels where the respective samples were collected, and set the temperature levels at which the algae have to grow in a large-scale production system [22, 45, 46]. Nutrient enrichment will be used to select species capable of growing in environments with high concentrations of nutrients. Algae growth that does not have these conditions has low biomass productivity and therefore is not suitable for commercial production. The temperature system will serve to speed up the process of determining the optimum temperature production of the species. If the selected algae do not grow, or grow slowly, in these conditions, these species are not suitable for commercial production [22, 45, 46].

Where the desired production environment is a saline environment, it will be necessary to select algae that present, in addition to the features already mentioned, salt tolerance for growth in environments with high salinity ranges [39, 45]. After these determinations, the pre-selection of species is done, mixing the species that presented the best conditions with the enrichment cultures. This step is performed especially when the culture medium is the open pond system (raceways) for the production of algal biomass [45].

These crops are grown in inoculated environments for long periods, allowing natural selection among species. Varieties are developed that are able to grow and compete in the best outdoor conditions in the environment in which they are grown for commercial production. Species that begin to adapt to these conditions have competitive power and environmental tolerance, both key features for reliable outdoor production on a large scale [45].

After these steps, the species already showing the temperature requirements, appropriate nutrients, and saline conditions, are isolated in separate cultures in order to explore their phenotypic differences (in particular with regard to lipid productivity) in the culture [45].

The species are then selected according to their lipid content by methods of selection and isolation of clones with high lipid content. After selecting this set of clonal cultures, they also undergo other selections regarding salt tolerance and performance at certain temperatures, in order to select the most productive strains for the culture medium. The species are then classified according to their general characteristics for large-scale cultivation [45].

The best lines are then cultured on the culture medium in smaller cultivation systems, in order to reproduce real behavior in various climatic cycles during the year. This allows the evaluation of performance in different environmental conditions, obtaining real productivity data. In this step it is also possible to evaluate the maintenance of contamination by other algae and protozoa. If the species perform well under these culture conditions and under conditions maximizing their productivity, then this species is chosen [45].

Thus, the criteria to be considered during the selection of species are: a) selection of the best strains in terms of lipid levels and high productivity, a better lipid profile, and adaptability to the type of water used and to environmental conditions; b) identification, in addition to biofuel production potential, of high-value-added compounds; c) selection of microalgae that have high photosynthetic efficiency; d) establishment of appropriate cultivation strategies that enable maximum productivity of lipids and biomass; e) ensuring priority in the search for organisms which have good productivity in wastewater and other extreme environments; f) identification of species highly tolerant to high levels of carbon dioxide (CO_2), contaminants and other physical effects; g) identification of the growing requirements of species, such as adding vitamins, minerals, *etc.;* h) performing lipid extraction and conversion into biodiesel using lower cost strategies [7, 22, 28, 38, 39].

Table **3** summarizes some of the main desirable properties in a microalgae species for large-scale cultivation in energy production [45].

Table 3. Summary of desirable properties of microalgae for large-scale culture for energy production.

Property	Reason
Rapid Growth	Required for high productivity (also reduces risk of contamination by other algae and/or predators and pathogens)
High lipid content (for biodiesel production)	Required for high lipid productivity
Optimum temperature	A high temperature optimum and tolerance (at least 30–35°C) is very important
High photosynthetic efficiency	Increases productivity
Ability to tolerate high irradiances	Reduces photoinhibition and photodamage at high irradiances and can increase productivity
Shear tolerance	Must tolerate the shear created by the circulation system
Growth in a "selective" environment	Makes management of contamination easier
Lower intrinsic respiratory rate	Higher productivity
Large cell size or colonial or filamentous morphology	Ease of harvesting
High specific gravity	Heavier cells are easier to harvest. However, heavier cells are harder to keep in suspension in culture system
Weak cell covering or no cell wall	Easier to extract product

(Table 3) contd.....

Property	Reason
No production of autoinhibitors	Reduces potential problems with medium recycling and allows productive high-density culture
Lipid composition (for biodiesel production)	Affects yield, efficiency of transesterification, and the quality of the biodiesel produced; Aside from taxonomic affiliation, lipid composition is affected by growth conditions (light, N-source, nutrient limitation, salinity, temperature *etc.*).

It is highly unlikely that a single species meets all the characteristics presented in Table **3**, thus the relative importance of each property will depend in part on the location and culture system.

Microalgae Cultivation

Microalgae culture presents three distinct components: a culture medium contained in a culture system; the algal cells growing on the medium; and air, allowing the carbon dioxide to circulate between the environment and the atmosphere [47].

Culture Medium

The production of lipids and fatty acid concentrations in microalgae are influenced mainly by the composition of the culture medium [7]. For the selection of the composition of the culture medium, one should take into account whether the type of algae to be used does not need the addition of a particular chemical substance to the culture medium (a positive growth rate being observed if added); or whether this should be done adding substances to the environment following known culture media methods, provided they do not bring negative effects on the growth rates of microalgae [47].

Knowledge of the characteristics of the natural habitat (eutrophic, oligotrophic) of the kind of microalgae in question is crucial for the choice of the culture medium [7, 47, 48].

Carbon, which is the main constituent of microalgae cells, can be obtained from inorganic sources (CO_2, $H_2CO_2^{-3}$, CO^{-3}), as well as from organic ones (hexoses, pentoses, *etc.*). Nitrogen (N) is also a very important element, because it is associated with the production of nucleic acids and proteins and directly related to the metabolism. These and other nutrients are maintained in culture medium at optimal intervals through known culture medium methods in order to promote the maximum growth [47].

The recipes of known culture medium are not always adequate for many species, and it is necessary to make trial and error tests for accurate determination of a culture medium for a given species. The culture media are thus classified as (1) defined media, which have all known constituents and can be attached to a chemical formula, and (2) undefined media, which consist of one or more natural or complex ingredients, whose composition is unknown [40, 47]. These culture media can be subdivided into freshwater culture media and saltwater culture media.

Freshwater culture media using composition recipes are widely used because they have characteristics similar to the natural habitats characteristic of various species of microalgae used in biomass production [47].

On the other hand, salt water is an ideal medium for growth of marine species, but it is a very complex medium containing dozens of unknown elements and a vast amount of varied organic compounds. Saltwater culture media, however, should be as simple as possible in its composition and preparation, comprising salt water (natural or artificial) as a base, which may be supplemented with various substances essential to the growth of microalgae such as nutrients, vitamins, chelating agents, and buffering compounds [47, 49].

Artificial defined media are often used as additives for natural media with unknown chemical characteristics in order to optimize them. They are widely used to simulate various nutritional and physical requirements of a particular species or group of species, especially when the exact nutritional needs are unknown [47, 49].

In addition to carbon dioxide (CO_2) and nitrogen (N), and the quality and availability of other nutrients and biological parameters, the productivity of lipids and growth of microalgae also depend on physical and chemical factors such as light, temperature, pH, salinity, and mixing of the culture medium [42].

Microalgae Production: Systems and Methods

Early research on cultivation of algae and their nutritional relations was traditionally dedicated to the cultivation of algae in their own natural environments, as well as their growth with respect to their ecosystem. In contrast, the culture system for commercial production requires maximum growth of cells having a high biomass density. For this, a better understanding of factors that intervene in the productivity of microalgae is needed, as well as the study of models of reactors capable of generating greater productivity at the lowest possible cost [48]. Several factors, such as need of light for photosynthetic growth, carbon dioxide, water, inorganic salts, as well as temperature control (within 20-30°C), make biomass production from microalgae generally more expensive than that of crops from other oil sources [28]. Therefore, the economic margins for the production of microalgal biomass that establish it as an economically viable source for biofuel production are much smaller, requiring a greater optimization of the microalgal biomass [48]. Several parameters should be utilized to reduce costs, such as: meeting the nutritional requirements of each species by providing the essential inorganic elements such as nitrogen (N), phosphorus (P), iron (Fe), and, in some cases, silicon (Si), as well as other minerals for vital structural and metabolic support of the biochemical cell; maintenance of pH for control, which is extremely important, since pH affects all biochemical aspects and cell ion absorption, thereby minimizing loss of carbon dioxide (CO_2); use of available CO_2 emissions by power plants, as well as the use of waste water; control of lighting levels, temperature, and growth temperature overnight [28, 48]. This optimization, however, focuses primarily on reducing costs and improving the efficiency of the methods and systems of microalgae culture. The cultivation methods can be performed in batch, semi-continuous, and continuous modes [44, 50]

Table **4** presents the advantages and disadvantages of microalgae cultivation methods [50].

Table 4. Advantages and disadvantages of microalgae cultivation methods.

Cultivation method	Advantages	Disadvantages
Batch	More used; simple; flexible	Less efficient; inconsistent quality
Semi-continuous	Easy; efficient	Sporadic and unpredictable quality
Continuous	Very efficient; cells with high quality; automation; higher production rate	Used only in smaller proportion cultures; complex; high cost of equipment

The batch method consists of a single inoculation of cells in the medium during a growth period of several days, until reaching the maximum/desired cell density. Then the algae is transferred to higher volume crop systems to continue its growth before it reaches its stationary phase, reaching its maximum cell density for harvesting. The batch culture method is widely used because of its simplicity and flexibility, allowing change of species in the medium and rapid repair of defects in the system. Despite being the most used, the batch method is not necessarily the most efficient, with limitations such as unpredictability of the quality of harvested cells, the need for prevention of contamination during inoculation and early growth, and the need for more manpower for harvest, cleaning, sterilization, and inoculation into the medium [7, 44, 50].

The semi-continuous method is a technique where there is a partial harvest of the biomass medium, followed by a new supply complementing the original volume and by a supplementing of nutrients for adjusting the initial level of enrichment. The other culture is then grown and harvested partially, and so on. Although it has unpredictable quality, the semi-continuous cultivation technique generally produces more than the batch technique, given the same size of culture medium [49, 50].

In the continuous method, there is the constant recharging of the culture medium in the system and, simultaneously, the constant removal of microalgae, keeping the volume constant. The continuous culture method can be divided into two categories, according to the cultivation controllers: turbidostat and chemostat. In a culture which uses turbidostat controllers, the concentration is maintained at a predetermined level. When the density reaches this predetermined level, a fresh medium is added to the culture in an automated system. In a culture that uses chemostat controlling, crop growth is controlled through the availability of nutrients in the medium. This method allows the maintenance of the cultures at very close to the maximum growth rate, producing algae with more predictability and with less need for work. The method, however, has limitations such as high costs and complexity [7, 44, 50].

These methods can be applied to both the open and closed cultivation systems, that is, in open reactors (open ponds, also called raceways) and closed reactors (photobioreactors). The feasibility of each system's techniques are influenced by the intrinsic properties of the species of algae to be used, costs, and climate conditions [28, 48, 51].

Open Pond Systems (Raceways)

Microalgae cultivation in open pond systems is so called because in this system the culture is directly exposed to the environment [45]. This system has been used for the production of microalgae in simple systems like natural lakes and ponds, and even in artificial ponds, since the 1950s [35]. Artificial open ponds are the main algae production systems, both commercially and for wastewater treatment, mainly because they are more economical on a large scale [45]. The most commonly used systems include shallow big ponds, tanks, circular ponds, and raceway ponds. Relevant aspects that must be considered in the production system design are: O_2 buildup generated to prevent photorespiration; the enrichment of the culture medium with CO_2 ; the limitation of lighting; ease of control of undesirable species and predators; other abiotic parameters which affect the culture medium, such as pH and salinity [45].

Among the open pond systems, raceway ponds are the most used. These ponds may be constructed of various types of materials, which determine the cost, performance, and durability of the ponds. Among the materials used are concrete, sand, and clay, and even more expensive materials such as PVC and glass fiber.

As illustrated, the broth containing nutrients and algae is introduced into the pond and circulated through mixing systems, while carbon dioxide (CO_2) is mechanically aerated in the system. Harvest occurs and the cycle restarts.

The advantages of open pond systems in microalgae biomass production on a large scale are these: they have less expensive costs when compared with closed photobioreactors; they do not compete with the food industry, and need no fertile and/or arable land, as they can be installed in areas with marginal production potential; they have lower maintenance costs [35, 45].

The open pond system suffers limitations due to natural weather conditions (rainy season) and inherent contamination and pollution threats from other algae, protozoa, and bacteria. Open pond systems thus require localization in selective environments [45].

The main limitation is low productivity (Table **5**) when compared to photobioreactors. This is due to environmental factors that, to some extent, cannot be controlled. Several factors result in low productivity in open pond systems: evaporation losses which result in changes in the ionic composition of the medium and hinder the growth of the crop; environmental conditions such as temperature changes and seasonal variations that cannot be controlled and directly affect system throughput; potential shortcomings due to the diffusion of CO_2 into the atmosphere, which can result in a decrease in biomass productivity due to less efficient use of CO_2 ; inefficient mixing mechanisms; and light limitation due to thickness of the upper layer [22, 35, 39, 48].

Table 5. Productivity in open pond systems, adapted from [35].

Microalgae species	Productivity (g/m² per day)
Spirulina platensis	14.0
Haematococcus pluvialis	15.1
Various species	19.0
Spirulina platensis	12.2
Spirulina platensis	19.4
Anabaena sp.	23.5
Chlorella sp.	23.5

(Table 5) contd.....

Microalgae species	Productivity (g/m² per day)
Chlorella sp.	11.1
Chlorella sp.	18.1

Although relatively high rates of algal biomass production may be achievable using open pond systems, according to data reported in the literature there is variability in the rates of production of the same species [35].

Closed Systems: Photobioreactors

The growing constraints of open pond systems, such as ease of contamination and low yields, have resulted in research and development regarding closed cultivation systems [28, 35, 39].

In the biofuels sector, much research focuses on cultivation using closed systems, because they have higher productivity of cell density compared to open pond systems.

Higher yield of biomass achieved with this type of system is due to the fact that photobioreactors allow greater control of biotic and abiotic growth parameters (temperature, *etc.*). They are also less prone to contamination and eliminate or drastically reduce evaporation. Photobioreactors also have the advantage of presenting flexible and favorable technical designs [45]. Aside from these advantages, the costs of installation and maintenance of closed systems are far superior to those of open pond systems [35].

The fundamental principle of photobioreactors is the reduction of the light path, increasing the amount of radiation received by each cell. Photobioreactors are more versatile than open pond systems, because they possess the ability to use light sources other than solar light, thereby providing the potential for increasing the photoperiod and improving on the variation of light intensity given by sunlight. However, the use of artificial light contributes to energy consumption, thereby increasing capital costs and impacting the sustainability of this system [7, 22].

Several designs and configurations of photobioreactors have been proposed, seeking to establish optimal conditions for obtaining maximum productivity, expected returns, and the system's viability. The most commonly used designs are the flat-plate, tubular, and column photobioreactors (Fig. **1**) [35].

Flat-plate photobioreactor systems are composed of transparent plates (glass or plastic), which may be positioned vertically or inclined, allowing maximum capture of solar radiation. Carbon dioxide (CO_2) and air are injected into the system by the base, to provide the carbon sources required for the development and growth of the biomass and to generate sufficient turbulence to expose the entire biomass to solar radiation [7, 35]. Photobioreactors are suitable for microalgae biomass cultivation, since this system provides a low accumulation of dissolved oxygen and high photosynthetic efficiency. This is due to the large surface area available for capturing light energy, offering great capacity for attaining high-density cells [22, 28]. However, this system has limitations, such as difficulties in industrial plants due to the height of the bioreactor, the possibility of growing biomass on the walls of the bioreactor, and difficulty in controlling the culture temperature [7].

Fig. (1). Closed photobioreactor systems: a) flat-plate; b) tubular [45].

Generally, tubular bioreactors are relatively cheaper compared to other closed systems. They are suitable for outdoor cultures because tubular bioreactors have large surface area illumination and exhibit good yields in biomass [35]. They consist of a transparent tube (plastic or glass) for collecting the sunlight. These tubular collectors have a limited diameter (0.1 meters or less), so that light penetrates sufficiently to the dense culture and thus ensures high biomass productivity [28]. The microalgae broth is dispensed from the reservoir (degassing column) which goes to the solar collectors, a fraction of the biomass is harvested, and the rest returns to the reservoir [28]. Some disadvantages of this system are the possibility of microalgae growth on the walls of the bioreactor, CO_2 depletion in the culture medium, accumulation of high concentrations of O_2, and marked variation in the pH [7, 28].

The photobioreactor column is a system composed of vertical columns where the air inlet occurs in the system base, aerating and mixing the culture. These systems are illuminated through transparent walls or with fluorescent lighting systems within the column, but the column systems have poor illuminated surface area due to their limited vertical form [7]. However, these systems have more efficient mixing, higher rates of gas-liquid transfer, and better conditions of biomass growth control. The culture thus undergoes less photoinhibition and photooxidation, guaranteeing a more appropriate light/dark cycle [7].

The hybrid system is a combination of open pond systems and closed systems (photobioreactors), addressing the benefits and limitations of both cultivation systems [52]. Table **6** shows the comparison of the open pond and closed photobioreactor cultivation systems [7].

Table 6. Comparison of different microalgae cultivation systems (open ponds and closed photobioreactors).

Microalgae cultivation systems	Closed photobioreactors	Open ponds
Contamination control	easy	difficult
Contamination risk	reduced	higher
Sterility	realizable	none
Process control	easy	difficult
Species control	easy	difficult
Mixing system	uniform	poor
Investment	high	low
Operating costs	high	low
Light use efficiency	higher	poor
Temperature control	more uniform temperature	difficult
Evaporation of the culture medium	low	high
Biomass concentration	3-5 times higher	smaller

Open ponds are very efficient for microalgae cultivation and are low cost systems. However, they very quickly present a high risk of contamination by unwanted species. On the other hand, photobioreactors are excellent for the maintenance of cultures with minimal possibility of contamination, but installation costs are much higher than in open pond systems [45].

A combination of systems is probably the most logical choice for profitable cultivation with high yield for biofuels production [48]. This promotes higher productivity as much for biomass as for lipids, since most microalgae species do not accumulate biomass and lipids simultaneously [52].

The system is implemented in two steps, the first working up the cell density, and the second working up lipids production. The first step takes place in a closed system (photobioreactor) under sufficient conditions of nutrients, in order to obtain the highest cell density possible with less risk of contamination. The inoculum produced is then transferred to the open ponds system under limited nutritional conditions, and microalgae begin to quickly convert the sun's energy into chemical energy stored as lipids [48]. The expected result is high biomass production with high lipid content [52].

Recovery of Microalgal Biomass

The recovery of microalgal biomass is one of the most challenging and problematic processes for industrial-scale processing systems of microalgae for biofuels production [34]. This problem occurs due to the microscopic size of microalgae cells (2-30 µm) and the large volumes of water in which they are grown, resulting in a low microalgal biomass density in the culture medium. This can reach, in exceptional cases, 5g dry cell weight (DW) per liter, but is usually around 0.3-0.5g dry cell weight per liter [22, 35].

The cost of recovery of the biomass from the broth is equivalent to about 20-30% of the total cost of production of biomass [53]. It is desirable to select a kind of microalgae that presents the favorable properties for recovery of biomass, such as greater cell size, high specific gravity in comparison with the environment, reliable self-flocculating, *etc.* [48]. However, in practice this can be achieved only in part, requiring continuing research and development of more efficient systems of recovery of biomass and with lower cost [45].

The selection of biomass recovery technology is crucial to economic performance in the production of microalgal biomass. Physical, chemical, or biological processes are used, and the key players are currently flocculation, filtration, flotation, gravity sedimentation, and centrifugation [48].

Generally, the process of solid-liquid separation involves two stages. In the first stage, the biomass is separated from the total volume of the culture medium, leading to a concentration factor of 100-800, depending on the initial concentration of biomass and the technologies used in the process. The main technologies employed in this stage are flocculation, flotation, or sedimentation by gravity [34]. In the second stage there is a ten to thirty fold increase of sludge concentration created in the first step. The use of technologies such as centrifugation, filtration, and ultrasonic aggregate cause intense use of energy at this stage [5, 34].

The search for appropriate technology and cost-effective recovery of biomass is an ongoing exercise where the techniques currently used have advantages and limitations. There is thus no method of recovery of biomass suitable for all production systems, because the technology selection and optimization of the process will depend as much on economic as on technical and operational factors.

Dewatering and Drying

The recovered biomass generates a sludge having a dry solids content of 5% to 15%. This material, which is perishable, must then undergo a dehydration and drying process, which converts it into a stable and storable product for further processing and enhances the extraction of intracellular components from the microalgae [35, 53 - 54].

By the use of draining tanks or the press, the dewatering process can increase the solids content of the microalgal biomass up to 20%. The biomass recovered may be directed to a tank where the water is settled and drained. The mechanical press expels water under pressure and, like the drain tank, directs the wastewater to a treatment facility if necessary and then returns it to the cultivation system. The only inputs used in the biomass dewatering process are the energy needed for operation and maintenance of the presses or drainage tanks [52].

In the drying process, the main methods that are used include sun drying, drying at low pressure, spray drying, freeze drying, fluidized bed drying, and drum drying. These methods differ in costs of investment and operation, as well as in the energy intensity required [35, 53, 55]. Drying method selection depends on the scale of production, as well as on the product to be dried [55].

Among the methods of dewatering and drying, sun drying is the most economically feasible method, but it has disadvantages which include long drying times, the requirement of large surface drying, and the risk of material loss. Spray drying is commonly used for the extraction of high-value-added products, but it is relatively expensive and can cause deterioration of biomass, especially in pigments. Freeze drying is also expensive, especially for large-scale operations, but facilitates the extraction of oil due to cell rupture [35, 55].

The drum dryer method for drying microalgal biomass is the most recommended. Due to the higher digestibility of the material produced, this method has lower energy requirements and lower maintenance costs [55].

Algal Lipid Extraction

Several types of lipids are produced by microalgae. Their distribution of these in the cells depends on the species as well as on cultivation conditions. Most lipids have fatty acids and can generally be classified into two categories: 1) neutral lipids comprising the glycerides (triglycerides, diglycerides, monoglycerides) and free fatty acids; 2) polar lipids that can be subcategorized as phospholipids and glycolipids. The neutral lipids are used in microalgae cells primarily for energy storage, while the polar lipids form the layers of cell membranes [56].

Some types of neutral lipids, which do not contain fatty acids such as hydrocarbons, ketones, sterols, pigments (carotenoids), and chlorophyll, are not convertible into biodiesel. Among the types of lipids extracted from microalgae, only triglycerides are easily converted into biodiesel by the transesterification method, which is the most common commercially adopted method [56, 57].

The composition and profile of fatty acids and lipids extracted from a particular species are also influenced by the harvest and separation of microalgae, the culture conditions, medium composition, extraction processes, and the characterization of the extracted lipid fraction. Given this context, the microalgal biomass lipid extraction method used for this purpose should be as specific and selective as possible, in order to minimize the extraction of non-lipid fractions and maximize the extraction of desired fractions [37, 42, 56].

Alternatively, non-extracted lipid fractions (proteins, polysaccharides, and pigments) can be utilized in the pharmaceutical and food industries, which may represent economic benefits aside from the production of biodiesel [37]. Procedures used in vegetable oil seeds, such as mechanical extraction, extraction by chemical solvents, and the combination of the two methods, have been adapted to microalgae [52]. Other methods have also been evaluated on a laboratory scale: supercritical fluid extraction; subcritical water extraction; enzyme extraction; ultrasound-assisted extraction; and osmotic shock [22, 37, 52]. The composition of lipids extracted from microalgae depends in part on the solvent used in their extraction [56]. Thus, the extraction technique is essential not only for the production process but also for determining the quality of the final product.

MICROALGAL BIODIESEL

The acceptability of biodiesel from microalgae as a fuel for use in fossil diesel engines is heavily dependent on compliance with existing standards. Generally speaking, biodiesel and petroleum diesel must contain similar fluid dynamics and thermodynamic properties, thus having equivalent characteristics, especially from the point of view of combustibility in diesel engines [27]. Biodiesel from microalgae must have physico-chemical properties that conform to international standards such as the International Biodiesel Standard for Vehicles (EN 14214) and the ASTM Biodiesel Standard.

Some authors have used *Chlorella protothecoides* species for the production of biodiesel from microalgae through the transesterification reaction. A crude lipid content of 55.2% microalgal biomass was reached [58]. The oil was extracted using the solvent n-hexane, and biodiesel was obtained by an acid catalyzed transesterification reaction.

Table **7** shows some of the properties required by international standards, comparing biodiesel from microalgae with first generation biodiesel and diesel fuel [58].

From Table **7**, it can be seen that the microalgae biodiesel has physical and chemical properties within the limits required by international standards (except for the kinematic viscosity at 40°C). Microalgae biodiesel also presented, in abundance, methyl esters of oleic acid, octadecadienoic acid methyl ester, and octadecanoic acid methyl ester, representing more than 80% of its composition and thus resulting in high quality biodiesel [58].

Table 7. Properties of diesel fuel, first generation biodiesel, biodiesel from microalgae, and limits set by international standards.

Property	Units	Diesel fuel	First generation biodiesel	Biodiesel microalgae	EN 14214	ASTM D6T51
Heating value	MJ/kg	40.0-45.0	31.8-42.3	41.0	-	-
Kinematic viscosity, 40°C	mm²/s	1.9-4.1	3.6-9.48	5.2	3.5-5.0	3.5/5.0
Specific gravity, 20° C	Kg/m³	838	860-890	864	860-900	860-900
Cetane number	-	-	45.0-65.0	-	> 51.0	-
Flash point	°C	75.0	100-170	115	> 101	> 100
Acid number	mg KOH/g	< 0.5	-	0.374	< 0.5	< 0.5

FINAL CONSIDERATIONS

The types of systems and technological methods proposed and employed in biodiesel production systems for microalgae biomass have undergone extensive research, and innovative technologies have been developed, such as hybrid systems, microalgal cell immobilization, supercritical/subcritical extraction, and *in situ* transesterification. The main objective is to create an efficient and sustainable system that reduces production steps, processing time, and the cost of the system, enabling the implementation of these large-scale technologies and thus enabling the production of biodiesel from microalgal biomass on an industrial scale.

As the energy productivity of microalgal biomass in biodiesel production presents high yields of biomass and lipid fractions, using a much smaller area of land, it may be concluded that microalgal biomass presents itself as a very good alternative due to the high content of lipids present in the cell. Furthermore, biomass derived from microalgae as a source of renewable biofuels, in contrast to other alternatives, has various characteristics, such as: it does not use arable land, and thus does not compete with the food industry; it does not require clean water; it also has the ability to mitigate environmental liabilities generated in other activities, with the possible association with other industries that produce liabilities that can serve as nutritional sources for the growth of microalgae (CO_2 and sugarcane vinasse, for example).

Microalgal biomass may also contain byproducts with high added-value that can be marketed in the food, animal feed, nutraceuticals, and cosmetics sectors, enabling the production of more than one product from the same biomass by increasing the gains in production.

Biodiesel from microalgae biomass presents physicochemical properties within the limits required by the international standards organization, such as ASTM D6751 and EN 14214, except for kinematic viscosity at 40°C. It also presents, in abundance, compounds that give high quality biodiesel microalgae.

Due to the high cost of the technologies employed, the high operating costs of maintenance and energy in modern production systems, and especially due to the low level of investment in research and in industrial plants for microalgae biodiesel productions, microalgae biomass is currently not a viable option for biofuel (biodiesel) production on a commercial scale. This is because it has a negative energy balance when compared to oil, or even to first generation biofuels (soy biodiesel, for example). However, it should be pointed out that breeding work on the selection and domestication of species and strains of microalgae is still being developed, as was done in the past with cultures of other raw materials such as soy, corn, *etc.* Accordingly, gains in productivity and reduction in production costs already attained for these other cultures have not yet been researched and developed with microalgae culture.

Microalgae is a raw material that is beginning to be more deeply researched as a renewable alternative for biofuel production. Given its sustainable qualities as compared to other raw materials, it is evident that microalgae should receive more attention and investment. After the development of genetic improvement and technological production systems, and if there really is an enventual reduction in production costs, microalgae can become a competitive alternative and a commercial reality in the biofuels industry.

CONFLICT OF INTEREST

The authors confirm that this chapter content has no conflict of interest.

ACKNOWLEDGEMENTS

None declared.

REFERENCES

[1]   Prusty BA, Chandra R, Azeez PA. Biodiesel: Freedom from Dependence on Fossil Fuels? Nature Proceedings , 2008 [28th Dec 2015]; Available at: http://precedings.nature.com/documents/2658/version/1/files/npre20082658-1.pdf

[2]   Martín M, Grossmann IE. On the Systematic Synthesis of Sustainable Biorefineries. Ind Eng Chem Res 2013; 52: 3044-64.

[3]   Nigam PS, Singh A. Production of liquid biofuels from renewable resources. Pror Energy Combust Sci 2011; 37: 52-68.

[4]   Sudarsan KG, Anupama PM. The relevance of biofuels. Curr Sci 2006; 90: 748-9.

[5]   Dragone G, Fernandes B, Vicente AA, Teixeira JA. Third Generation Biofuels from Microalgae; Current Research, Technology and Education Topics in Applied Microbiology and Microbial Biotechnology [28th Dec 2015]; Available at: http://www.formatex.info/microbiology2/1355-1366.pdf

[6]   Kagan J. Third and Fourth Generation Biofuels: Technologies, Markets and Economics Through; Greentech Market Research , [28th Dec 2015]; Available at: https://www.greentechmedia.com/research/report/third-and-fourth-generation-biofuels

[7]   Mata TM, Martins AA, Caetano NS. Microalgae for biodiesel production and other applications: A review. Renew Sustain Energy Rev 2010; 14: 217-32.

[8]   Parente EJ. Biodiesel: uma aventura tecnológica num país engraçado. 1st ed., Fortaleza: Unigráfica 2003.

[9]   Knothe G. Manual de Biodiesel. 1st ed., São Paulo: Edgard Blücher 2006.

[10]   US Energy Information Administration. Annual Energy Outlook 2014 with projections to 2040 , 2014 [28th Dec 2015]; Available at: http://www.eia.gov/forecasts/aeo/pdf/0383(2014).pdf

[11]   Freitas SM, Fredo CE. Fontes energéticas e protocolo de Kyoto: a posição do Brasil. Informações Econômicas 2005; 35: 77-82.

[12]   Lima Filho DO, Sobage VP, Calarge TC. Mercado do biodiesel: um panorama mundial. Revista Espacios 2008; 29: 2-2.

[13]   BP GLOBAL. Statistical Review of World Energy , 2014 [28th Dec 2015]; Available at: http://www.eia.gov/forecasts/aeo/pdf/0383(2014).pdf

[14]   Plá JA. Histórico do Biodiesel e suas Perspectivas , [28th Dec 2015]; Available at: http://www.ufrgs.br/decon/publionline/textosprofessores/pla/hist_rico.doc

[15]   Guo M, Song W, Buhain J. Bioenergy and Biofuels: History, status, and perspective. Renew Sustain Energy Rev 2015; 42: 712-25.

[16]   Khalil CN. As tecnologias de produção de biodiesel. O Futuro da Indústria: Biodiesel, Coletânea de Artigos Brasília DF: Ministério do Desenvolvimento, Indústria e Comércio Exterior , [28th Dec 2015]; Available at: http://www.desenvolvimento.gov.br//arquivos/dwnl_1201279825.pdf

[17]   Geris R, Santos NA, Amaral BA, Maia IS, Castro VD, Carvalho JR. Biodiesel de soja – Reação de transesterificação para aulas práticas de química orgânica. Quim Nova 2007; 30: 1369-73.

[18]   Ramos LP, Brugnagoa RJ, Silva FR, Cordeiro CS, Wypych F. Esterificação e transesterificação simultâneas de óleos ácidos utilizando carboxilatos lamelares de zinco como catalisadores bifuncionais. Quim Nova 2014; 38: 46-54.

[19]   Ferreira SP, Souza-Soares L, Costa JA. Revisão: microalgas: uma fonte alternativa na obtenção de ácidos gordos essenciais. Rev Ciênc Agrár (Belém) 2013; 36: 275-87.

[20]   Gui MM, Lee KT, Bhatia S. Supercritical ethanol technology for the production of biodiesel: Process optimization studies. J Supercrit Fluids 2009; 49: 286-92.

[21]   Garcia CM. Master *Dissertation.Transesterificação* de óleos vegetais. Campinas: University of Campinas. 2006.

[22]   Rawat I, Kumar RR, Mutanda T, Bu F. Biodiesel from microalgae: A critical evaluation from laboratory to large scale production. Appl Energy 2013; 103: 444-67.

[23]   Huang GH, Chen F, Wei D, Zhang XW, Chen G. Biodiesel production by microalgal biotechnology. Appl Energy 2010; 87: 38-46.

[24]   Canozzi JB. Master Dissertation Avaliação do extrato da semente do maracujá (Passiflora edulis Sims) obtido por extração com dióxido de carbono supercrítico: influência do emprego de co-solvente e possíveis impactos da aplicação no processo de transesterificação com fluidos supercríticos Universidade Federal de Santa Catarina 2013.

[25]   Stamenkovic OS, Velickovic AV, Velijkovic B. The production of biodiesel from vegetable oils by ethanolysis: Current state and perspectives. Fuel 2011; 90: 3141-55.

[26]   Haas MJ, Foglia TA. Matérias-primas alternativas e tecnologias para a produção de biodiesel. Manual de Biodiesel. 1st ed. São Paulo: Edgard Blücher 2006; pp. 29-45.

[27]   Parente EJ. Biodiesel no plural , [28th Dec 2015]; Available at: http://www.desenvolvimento. gov.br//arquivos/dwnl_1201279825.pdf

[28]   Chisti Y. Biodiesel from microalgae. Biotechnol Adv 2007; 25: 294-306.

[29]   Deng X, Li Y, Fei X. Microalgae: A promising feedstock for biodiesel. Afr J Microbiol Res 2009; 3: 1008-14.

[30]   Richmond A. Handbook of Microalgal Culture: Biotechnology and Applied Phycology. 1st ed., Lowa: Blackwell Publishing 2004.

[31]   Li Y, Horsman M, Wang B, Wu N, Lan CQ. Effects of nitrogen sources on cell growth and lipid accumulation of green alga *Neochloris oleoabundans*. Appl Microbiol Biotechnol 2008; 81: 629-36.

[32]   Lourenço SO. Cultivo de Microalgas Marinhas – Princípios e Aplicações. 1st ed., São Carlos: RiMa 2006.

[33]   Kuamar A, Ergas S, Yuan X, *et al.* Enhanced CO_2 fixation and biofuel production *via* microalgae: recent developments and future directions. Trends Biotechnol 2010; 28: 371-80.

[34]   Wang B, Li Y, Wu N, Lan CQ. CO_2 bio-mitigation using microalgae. Appl Microbiol Biotechnol 2008; 79: 707-18.

[35]   Brennan L, Owende P. Biofuels from microalgae - A review of technologies for production, processing, and extractions of biofuels and co-products. Renew Sustain Energy Rev 2010; 14: 557-77.

[36]   Suarez P, Meneghetti SM, Meneghetti MR, Wolf CR. Transformação de triglicerídeos em combustíveis, materiais poliméricos e insumos químicos: algumas aplicações da catálise na oleoquímica. Quim Nova 2007; 30: 667-76.

[37]   Franco AL, Lôbo IP, Cruz RS, Teixeira CM, Neto JA, Menezes RS. Biodiesel de microalgas: avanços e desafios. Quim Nova 2013; 36: 437-48.

[38]   Pereira CM, Hobuss CB, Maciel JV, *et al.* Biodiesel renovável derivado de microalgas: avanços e perspectivas tecnológicas. Quim Nova 2012; 35: 2013-8.

[39]   Sheehan J, Dunahay T, Benemann J, Roessler P. Look Back at the US Department of Energy's Aquatic Species Program - Biodiesel from Algae National Renewable Energy Laboratory. 1st ed., Colorado: National Renewable Energy Laboratory 1998.

[40]   Derner RB, Ohse S, Villela M, Carvalho SM, Fett R. Microalgas, produtos e aplicações. Cienc Rural 2006; 36: 1959-67.

[41]   Ahmad AL, Yasin NH, Derek CJ, Lim JK. Microalgae as a sustainable energy source for biodiesel production: A review. Renew Sustain Energy Rev 2011; 15: 584-93.

[42]   Scott SA, Davey MP, Dennis JS, *et al.* Biodiesel from algae: challenges and prospects. Curr Opin Biotechnol 2010; 21: 277-86.

[43]   Kleinová A, Cvengrošová Z, Rimarčík J, Buzetzki E, Mikulec J, Cvengroš J. Biofuels from algae. Procedia Eng 2012; 42: 231-8.

[44]   Suali E, Sarbatly R. Conversion of microalgae to biofuel. Renew Sustain Energy Rev 2012; 16: 4316-42.

[45]   Borowitzka MA. Energy from Microalgae: A Short History. Develop Appl Phycol 2013; 5: 1-15.

[46]   Siver PA. A new thermal gradient device for culturing algae. Br Phycol J 1983; 18: 159-64.

[47]   Barsanti L, Gualtieri P. Algae Anatomy, Biochemistry, and Biotechnology. 1st ed., Boca Raton, Florida: CRC Press 2006.

[48]   Schenk PM, Thomas-Hall SR, Stephens E, *et al.* Second Generation Biofuels: High-Efficiency Microalgae for Biodiesel Production. BioEnergy Res 2008; 1: 20-43.

[49]   Lavens P, Sorgeloos P. Manual on the production and use of live food for aquaculture. FAO Fisheries Technical Paper , 1996 [28th Dec 2015]; Available at: ftp://ftp.fao.org/docrep/fao/003/w3732e/w3732e00.pdf

[50]   Ghasemi Y, Rasoul-Amini S, Naseri AT, Montazeri-Najafabady N, Mobasher MA, Dabbagh F. Microalgae biofuel potentials. Appl Biochem Microbiol 2012; 48: 150-68.

[51]   Benemann Y, Oswald WJ. Systems and economic analysis of microalgae ponds for conversion of CO2 to biomass. Final report SciTech Connect , 2003 [28th Dec 2015]; Available at: http://www.osti.gov/scitech/servlets/purl/493389

[52]   Ryan C. Cultivating Clean Energy: The Promise of Algae Biofuels. NRDC Natural Resources Defense Council , [28th Dec 2015]; 2009 Available at: http://www.nrdc.org/energy/cultivating.asp

[53]   Grima EM, Belarbi EH, Fernández FG, Robles Medina A, Chisti Y. Recovery of microalgal biomass and metabolites: process options and economics. Biotechnol Adv 2003; 20: 491-515.

[54]   Surendhiran D, Vijay M. Microalgal Biodiesel - A Comprehensive Review on the Potential and Alternative Biofuel. Res J Chem Sci 2012; 2: 71-82.

[55]   Shelef G, Sukenik A, Green M. Microalgae Harvesting and Processing: A Literature Review Golden, Colorado: US Department of Energy , [28th Dec 2015];1984 Available at: http://www.nrel.gov/docs/legosti/old/2396.pdf

[56]   Halim R, Danquah MK, Webley PA. Extraction of oil from microalgae for biodiesel production: A review. Biotechnol Adv 2012; 30: 709-32.

[57]   Bai M, Cheng C, Wan H, Lin Y. Microalgal pigments potential as byproducts in lipid production. J Taiwan Institute of Chemical Engineers 2011; 42: 783-6.

[58]   Xu H, Miao X, Wu Q. High quality biodiesel production from a microalga Chlorella prototothecoides by heterotrophic growth in fermenters. J Biotechnol 2006; 126: 499-507.

SUBJECT INDEX

A

Accelerated breeding 61, 72, 76, 126
Advances in cultivation technologies i, 142, 144
Agricultural productivity 142
Agricultural wastes 124
Agroinfiltration 83, 85, 91, 92
Alfalfa 124
Algal lipid extraction 192
Anther culture 18-19
AP endonuclease 46, 53
AP endonuclease assay 53

B

Base excision repair (BER) 44, 46, 47
Beet 104, 124, 125
BER pathway 43, 46, 48, 53, 54
Biodiesel from microalgae 169, 176, 193-195
Biodiesel history 171
Biodiesel in the world 171
Biodiesel production 29, 108, 110-111, 113, 124, 169-170, 172, 174-178, 180, 182-183, 194-195
Bioenergy crops ii, 103, 105-106, 123, 125-127, 131, 142
Bioenergy species 126
Bioethanol 16, 94, 108, 111, 124, 127, 148, 170
Biofuel crops i, 111, 122
Bioinformatics 122, 123, 128, 130-135
Bioinformatics tools 122, 123, 128, 130, 134
Biomass accumulation 142-144, 149, 153
Biomass production for biofuels 142
Biomass yield i, 106, 107, 142, 144
Bioreactors 1, 28-29, 189

C

Callus 5, 16, 19, 22, 25, 27-28, 30
Canola 76, 124-125, 177
Cassava 6, 9, 125
Cell suspension culture 15, 16, 27-29
Challenges in biomass production 142

Cisgenesis 70-71
Citrus sp. 19, 68, 72
Computational analysis of molecular sequences 129
Corn i, 2, 46, 104, 107-108, 124-125, 171-172, 177, 195
CRISPR/*Cas 9* 62, 66
Crocus sp. 10
Cultivation technologies i, 142, 144
Culture medium 7-10, 19-20, 27, 33, 165, 181, 183-184, 186, 189-191

D

2D gel electrophoresis 103
Dewatering and drying 192
Differential expression analyses 130
Direct reversal repair (DRR) 44-45
DNA damage 43-45, 49, 55
DNA glycosylase 46, 48, 53
DNA glycosylase assay 53
DNA repair 43-46, 48, 50-51, 55
DNA sequencing platforms 128
DRR pathway 43

E

Elephant grass 124
Electrophoretic karyotyping 162, 165-167
Energy cane 104, 107, 142, 151-152
Eucalyptus micropropagation 7
Eucalyptus 6, 7, 104, 109, 123, 124, 126, 131, 133

F

Factors influencing biomass yield i, 142, 144
Feedstock i, 94, 102, 104, 108, 109, 113, 142, 150, 163, 178
Fermentation i, 1, 29, 94, 103-104, 108, 112, 113, 114, 125, 151, 162-167
Fingerprinting 162, 165-166
First generation ethanol 104, 143, 148
Flocculation 167, 191

www.ingramcontent.com/pod-product-compliance
Lightning Source LLC
Chambersburg PA
CBHW041727210326
41598CB00008B/800